# Engineering Materials

This series provides topical information on innovative, structural and functional materials and composites with applications in optical, electrical, mechanical, civil, aeronautical, medical, bio- and nano-engineering. The individual volumes are complete, comprehensive monographs covering the structure, properties, manufacturing process and applications of these materials. This multidisciplinary series is devoted to professionals, students and all those interested in the latest developments in the Materials Science field, that look for a carefully selected collection of high quality review articles on their respective field of expertise.

**Indexed at Compendex (2021) and Scopus (2022)**

Hairus Abdullah • Januar Widakdo
Editors

# Membrane Physics of Carbon-Based Materials

## Fundamentals, Fabrication, and Applications

Springer

*Editors*
Hairus Abdullah
Department of Materials Science and Engineering
National Taiwan University of Science and Technology
Taipei, Taiwan

Januar Widakdo
Department of Physics
Universitas Indonesia
Depok, Indonesia

ISSN 1612-1317 ISSN 1868-1212 (electronic)
Engineering Materials
ISBN 978-981-95-5604-5 ISBN 978-981-95-5605-2 (eBook)
https://doi.org/10.1007/978-981-95-5605-2



This Springer imprint is published by the registered company Springer Nature Singapore Pte Ltd.
The registered company address is: 152 Beach Road, #21-01/04 Gateway East, Singapore 189721, Singapore

# Preface

Separation processes are fundamental to modern industry, underpinning everything from water purification and clean energy production to chemical manufacturing. Yet, many of these processes remain energy-intensive, creating an urgent need for more efficient, durable, and selective materials. Carbon-based membranes have emerged as a leading candidate to meet this challenge, offering a unique combination of molecular precision, remarkable stability, and versatile functionality. This book, *Membrane Physics of Carbon-Based Materials: Fundamentals, Fabrication, and Applications*, provides a timely and comprehensive exploration of this dynamic field. It is structured to guide the reader on a logical journey from fundamental principles to future frontiers.

The journey begins with chapter "Fundamental Aspects and Evolution of Carbon-Based Membranes", which lays a solid foundation by reviewing the historical evolution and intrinsic properties of carbon membranes. It establishes why materials like carbon nanotubes, graphene, and carbon molecular sieves are so compelling, while also honestly addressing the inherent challenges of fragility and scalability. Building on this foundation, chapter "Innovative Synthesis Approaches for High-Performance Carbon-Based Membranes" delves into the heart of materials engineering: synthesis. It provides a detailed comparison of advanced fabrication techniques from pyrolysis to chemical vapor deposition, illuminating how precise control over the synthesis process dictates the critical parameters of pore structure, morphology, and ultimately, membrane performance. With the "how" of creation established, chapter "Understanding Mass Transport in Carbon-Based Membranes" explores the "why" of function by examining the fundamental mechanisms of mass transport. This chapter connects the intricate pore architecture, crafted through synthesis, to the practical performance of the membrane, explaining how gases and liquids move at the molecular level and how this movement can be modeled and optimized. Chapter "Carbon-Based Membranes for Sustainable Energy and Environmental Applications" then bridges the gap from theory to practice, showcasing the real-world impact of these materials. It surveys the expansive applications of carbon membranes in addressing global challenges, particularly in gas separation for $CO_2$ capture and advanced water purification, highlighting the role of innovative composites and computational tools. Finally, chapter "Advancing Carbon-Based Membrane Technologies for the Future" looks ahead, offering a critical perspective on the future. It identifies the key frontiers from AI-driven molecular design to the engineering challenges of commercial-scale manufacturing that will define the next generation of carbon membrane technology and its pathway to industrial adoption.

This volume consolidates the collective knowledge of a rapidly advancing field, serving as an essential resource for students, researchers, and engineers in materials science, chemical engineering, and environmental technology. It not only captures the current state-of-the-art but also inspires the innovation required to overcome existing limitations. We invite you to explore the pages that follow and discover the immense potential of carbon-based membranes to shape a more sustainable and efficient future.

Taipei, Taiwan
Depok, Indonesia
November 2nd, 2025

Hairus Abdullah
Januar Widakdo

# Contents

# Contributors

**Hairus Abdullah** Department of Industrial Engineering, Universitas Prima Indonesia, Medan, Indonesia

**Fena Kharisma Annisa** Department of Physics, Faculty of Mathematics and Natural Sciences, Universitas Indonesia, Depok, Indonesia

**Nur Halimah** Department of Physics, Faculty of Mathematics and Natural Sciences, Universitas Indonesia, Depok, Indonesia

**Navilah Auva Hanandara** Department of Physics, Faculty of Mathematics and Natural Sciences, Universitas Indonesia, Depok, Indonesia

**Muhammad Daffa Nuur Hidayat** Department of Physics, Faculty of Mathematics and Natural Sciences, Universitas Indonesia, Depok, Indonesia

**Muhammad Arrayyan Pasha Marseno** Department of Physics, Faculty of Mathematics and Natural Sciences, Universitas Indonesia, Depok, Indonesia

**Haidah Novianti** Department of Physics, Faculty of Mathematics and Natural Sciences, Universitas Indonesia, Depok, Indonesia

**M. David Domini Oppusunggu** Department of Physics, Faculty of Mathematics and Natural Sciences, Universitas Indonesia, Depok, Indonesia

**Yulia Mariana Tesa Ayudia Putri** Department of Chemistry, Faculty of Mathematics and Natural Sciences, Universitas Indonesia, Depok, Indonesia

**Isnaini Rahmawati** Department of Chemistry, Faculty of Mathematics and Natural Sciences, Universitas Indonesia, Depok, Indonesia

**Aghniya Erytama Ramadhan** Department of Physics, Faculty of Mathematics and Natural Sciences, Universitas Indonesia, Depok, Indonesia

**Muhammad Ryan Afrizal Ramadhani** Department of Physics, Faculty of Mathematics and Natural Sciences, Universitas Indonesia, Depok, Indonesia

**Aditya Rianjanu** Department of Materials Engineering, Institut Teknologi Sumatera, Terusan Ryacudu, Way Hui, Jati Agung, Lampung, Selatan, Indonesia

Center for Green and Sustainable Materials, Institut Teknologi Sumatera, Terusan Ryacudu, Way Hui, Jati Agung, Lampung, Selatan, Indonesia

**Gavin Athaya Sabian** Department of Physics, Faculty of Mathematics and Natural Sciences, Universitas Indonesia, Depok, Indonesia

**Shabira** Department of Physics, Faculty of Mathematics and Natural Sciences, Universitas Indonesia, Depok, Indonesia

**Ferdy Subandi** Department of Physics, Faculty of Mathematics and Natural Sciences, Universitas Indonesia, Depok, Indonesia

**Adhiyoga Perdananajati Sulistiono** Department of Physics, Faculty of Mathematics and Natural Sciences, Universitas Indonesia, Depok, Indonesia

**Abraham Nathan Sunanta** Department of Physics, Faculty of Mathematics and Natural Sciences, Universitas Indonesia, Depok, Indonesia

**Januar Widakdo** Department of Physics, Faculty of Mathematics and Natural Sciences, Universitas Indonesia, Depok, Indonesia

**Abisha Jaumar Widjanarko** Department of Physics, Faculty of Mathematics and Natural Sciences, Universitas Indonesia, Depok, Indonesia

# Fundamental Aspects and Evolution of Carbon-Based Membranes

**Gavin Athaya Sabian, Adhiyoga Perdananajati Sulistiono, Muhammad Ryan Afrizal Ramadhani, and Januar Widakdo**

## 1 Introduction

In recent years, membrane technology has become a vital component of various industrial and research separation applications due to its energy efficiency, scalability, and versatility [1]. Among the numerous materials studied for membrane development, carbon-based membranes have gained considerable attention over the past few decades [2]. Initially developed in the 1970s for gas separation, carbon membranes have since evolved into a wide array of separation media, now including carbon allotropes like carbon molecular sieves, carbon nanotubes, and graphene derivatives [3]. This rapid advancement has been driven by the remarkable thermal and chemical stability of carbon [4]. Its tunable porosity and surface chemistry enable selective transport across a range of applications, including gas separation [5–10], water purification [11–14], and solvent nanofiltration [15–18]. Research into carbon membrane structures has also opened new avenues for permeability and selectivity that surpass the limitations of traditional polymeric membranes [19]. As membrane science advances into the nanotechnology era, it is paramount to understand the historical development, material diversity, and future potential of carbon-based membranes to guide further innovation and application [20].

G. A. Sabian · A. P. Sulistiono · M. R. A. Ramadhani · J. Widakdo (✉)
Department of Physics, Faculty of Mathematics and Natural Sciences, Universitas Indonesia, Depok, Indonesia
e-mail: januar.widakdo@ui.ac.id



H. Abdullah, J. Widakdo (eds.), *Membrane Physics of Carbon-Based Materials*, Engineering Materials, https://doi.org/10.1007/978-981-95-5605-2_1

## 2 History and Development of Carbon-Based Membranes in Various Applications

The concept for carbon membrane or film for gas separation was found in the early 1970. Ash, Barrerm, & Lowson had compressed non-porous graphited carbon into a plug, called carbon membrane [21]. Significant progress was made when Koresh and Soffer made carbon molecular sieves produced from the pyrolysis of polymeric materials. The results have proved to be very ideal for gas separation through molecular sieving, particularly in systems like oxygen/nitrogen or hydrogen/methane separations [22].

Later in the 1990s, carbon membranes began to attract attention for their improved thermal stability and chemical resistance compared to polymeric membranes, with an emphasis in harsh industrial environments. Researchers studied a range of synthetic precursors including polyimides, phenolic resins, and pitch-based materials to fine-tune pore structure and morphology during carbonization [23, 24]. Moreover, Sumio Iijima's 1991 article about his discovery in carbon nanotubes (CNT) gave rise to huge interests in carbon nanostructures [25]. Along with advancements in nanotechnology, this breakthrough laid the foundation for extensive research into their unique physical properties.

In the 2000s, progress in nanotechnology introduced a new generation of carbon-based materials to the world, such as carbon nanotubes (CNTs), graphene, and activated carbon nanocomposites. CNT-based membranes gained traction for their unique property to facilitate ultrafast and selective transport of gases and liquids due to their atomically smooth inner walls and hydrophobic confinement effects [26, 27]. The subject of The Nobel Prize in Physics 2010, Graphene, is also widely researched. Graphene is prominent for as a unique strong yet thin, transparent, and flexible conductor [28, 29]. Graphene oxide (GO) membranes also emerged as promising platforms for water purification, leveraging the tunability of interlayer spacing to sieve molecules with high precision [30].

In recent years, research has focused on integrating carbon materials with hybrid architectures. This includes polymer-carbon composites and carbon bilayers that improves mechanical strength and scalability. Advancements in additive manufacturing and advanced pyrolysis techniques have facilitated better control over membrane thickness, defect density, and pore size [31]. Meanwhile, the rise of sustainable engineering has sparked massive research efforts into biomass-derived activated carbons as membrane precursors. These research efforts aim to reduce reliance on synthetic polymers and improve lifecycle sustainability [32].

Due to persistent fabrication challenges, fouling sensitivity, and competition from more mature polymeric membrane technologies, commercial efforts of carbon membranes remain limited despite decades of development. However, their exceptional performance under extreme thermal and chemical environments ensures that carbon membranes continue to play an important role in applications such as high-temperature hydrogen recovery, $CO_2$ capture from flue gas [5–10], and organic solvent nanofiltration [33–35].

# 3 Types of Carbon Materials in Membrane Technology

The development of carbon-based membranes has sparked the exploration and utilization of a diverse collection of carbon materials with differing and unique structural characteristics, fabrication techniques, and separation performances [2, 36]. The most widely used carbon materials in membrane technology include activated carbon [37], carbon molecular sieves (CMS) [38], carbon nanotubes (CNTs) [39], graphene and its derivatives [40], carbide-derived carbons (CDCs) [41], and carbon aerogels [42]. These materials are primarily used for gas separation [5–10], water purification [11–14], and organic solvent nanofiltration [15–18].

## *3.1 Activated Carbon and Activated Carbon Cloth*

Activated Carbon (AC) represents one of the earliest and most widely adopted forms of engineered carbon materials, prominently known for its exceptionally high surface area and intrinsic porosity [43]. Due to these attributes, AC is typically employed in membrane technology, often integrated in powder or woven cloth forms (Activated Carbon Cloth, or ACC), serving primarily as an efficient adsorptive layer [44]. Its notable ion storage capacity and excellent conductivity have made it a staple in electrochemical systems, particularly Capacitive Deionization (CDI) systems for water treatment [45]. The operational principle of CDI relies on applying a voltage difference across porous carbon electrodes to electrostatically attract and store ions, thereby producing desalinated water. This mechanism is schematically presented in Fig. 1D, which also contrasts standard CDI with the enhanced Membrane Capacitive Deionization (MCDI), where ion-exchange membranes are used to boost efficiency and selectivity [45].

However, despite its low cost and ready availability, Activated Carbon suffers from lower selectivity when compared to more advanced, precisely engineered carbon materials like Carbon Molecular Sieve Membranes (CMSMs). The structural complexity of AC is critical: Fig. 1B illustrates this by contrasting the typically microporous network of Activated Carbon Fiber (ACF) with the broader distribution of micro-, meso-, and macropores found in granular AC. Modern research often seeks to enhance AC's functionality through structural hybridization, such as growing carbon nanofibers (CNFs) directly onto activated carbon fiber fabrics (ACFFs). The morphological results of this enhancement, both before and after acid treatment, are shown via SEM images in Fig. 1A. Beyond conventional filtration, carbon's versatile properties are leveraged in emerging applications like solar energy; for example, Fig. 1C presents a schematic of a solar steam generation device where Activated Carbon Fiber Cloth (ACFC) and CNFs act as a photothermal material, facilitating highly efficient water evaporation while managing salt rejection through regulated transport paths.

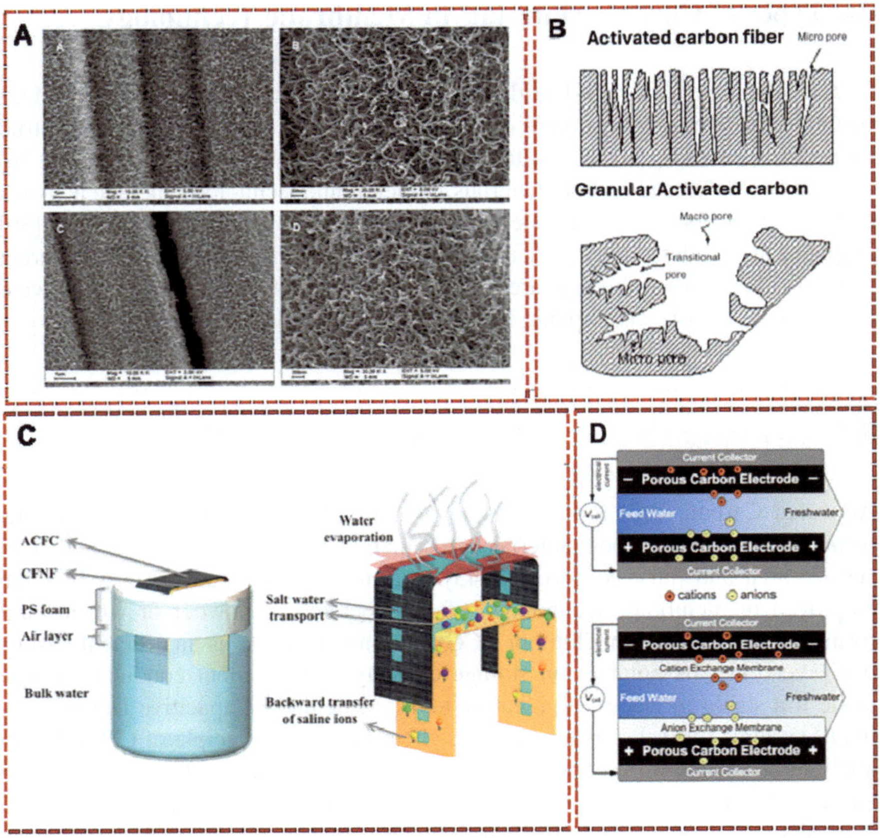


**Fig. 1** (**A**) shows scanning electron microscope (SEM) images of carbon nanofibers (CNFs) grown on activated carbon fiber fabrics (ACFFs). Images (**a**) and (**b**) depict the as-grown CNFs, while images (**c**) and (**d**) show the CNFs after acid treatment [46]. Reproduced from ref. [46] with permission from Elsevier, copyright 2006. (**B**) illustrates the structural differences between activated carbon fiber (ACF) and granular activated carbon. ACF is shown with a microporous structure, whereas granular activated carbon displays a more varied pore network including micropores, transitional pores, and macropores [47]. Reproduced from ref. [47] with permission from Elsevier, copyright 2006. (**C**) presents a schematic of a solar steam generation device utilizing carbon-based materials. The device design (a) includes an activated carbon fiber cloth (ACFC) and carbon nanofibers (CNF). The accompanying diagram (b) illustrates the transport paths for water and salt, showing how water evaporates while saline ions are transferred backward [48]. Reproduced from ref. [48] with permission from Elsevier, copyright 2019. (**D**) is a schematic of a cell for capacitive deionization (CDI). It demonstrates how a voltage difference applied to porous carbon electrodes attracts cations and anions, leading to the production of desalinated water. The figure contrasts standard CDI with membrane capacitive deionization (MCDI), where ion-exchange membranes are used to enhance the process [45]. Reproduced from ref. [45] with permission from Elsevier, copyright 2013

## 3.2 Carbon Molecular Sieves (CMSs)

Carbon Molecular Sieves (CMSs) represent a critical advancement in high-performance separation materials, generated through the pyrolysis (carbonization) of suitable polymeric precursors such as polyimides and phenolic resins [49]. The material's exceptional capability stems from its highly rigid structure, which possesses a network of ultramicropores (typically ~3–7 Å) ideal for molecular sieving [50, 51]. These narrow pores are formed precisely during carbonization and are capable of differentiating molecules based on subtle differences in their kinetic diameter. This mechanism makes CMS membranes highly effective in crucial industrial separations, notably $H_2/CH_4$ and $O_2/N_2$, offering the added advantage of excellent thermal and chemical stability under harsh operating conditions [52, 53].

Current research focuses heavily on precise control over the synthesis process to maximize performance. Figure 2A illustrates an advanced preparation technique for a hydrogen-bonded organic framework-derived carbon molecular sieve (HCMS) membrane, detailing the stages from precursor solution via spin-coating to the final carbonization process. This highlights the foundational role of the precursor and thermal treatment in defining the final CMS structure. The complexity of this structure is further detailed in Fig. 2B, where panel (a) shows the single-layer carbon units, contrasting the $sp^3$-C structure in CMS–600 with the predominantly $sp^2$-C structure in CMS–750. Panel (b) provides three-dimensional CMS models, and panel (c) uses a simulated pore morphology to visually distinguish pore volume (blue) from atomic volume (red), demonstrating how pyrolysis conditions create varied pore structures.

Furthermore, CMS materials are frequently incorporated into Mixed-Matrix Membranes (MMMs) to combine their high selectivity with the mechanical stability of a polymer matrix. Figure 2C provides a schematic of the gas transport mechanism within a CMS/ZSM–5 MMM. This diagram shows various gas molecules ($H_2$, $CO_2$, $O_2$, $N_2$, $CH_4$) navigating the composite material, specifically indicating how the ZSM–5 particles and the interface gap influence the selective transport. Ultimately, tuning the CMS pore structure—either through precise carbonization parameters or post-treatment techniques—remains the key strategy for significantly enhancing both the selectivity and permeance of these advanced membranes.

## 3.3 Carbon Nanotubes (CNTs)

Aligned Carbon Nanotube (CNT) membranes represent a distinct frontier in separation technology, offering a unique pathway for fluid and gas transport fundamentally different from traditional porous media [57]. Their atomically smooth and often hydrophobic inner walls induce nanoscale confinement and facilitate slip flow mechanisms [26, 27], resulting in rapid and highly selective transport. These characteristics make CNTs particularly appealing for applications requiring ultrafast

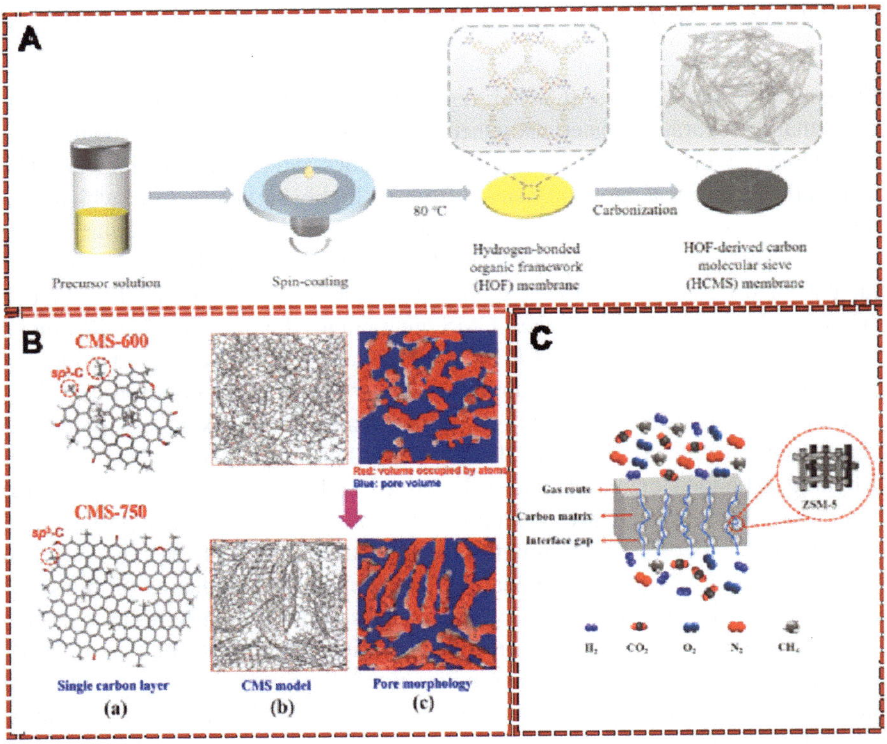


**Fig. 2** (**A**) shows the preparation process for a hydrogen-bonded organic framework-derived carbon molecular sieve (HCMS) membrane. The process involves preparing a precursor solution, which is then applied via spin-coating at 80 °C to form a hydrogen-bonded organic framework (HOF) membrane. Finally, the HOF membrane undergoes carbonization to produce the HCMS membrane [54]. Reproduced from ref. with permission [54] from Elsevier, copyright 2023. (**B**) depicts the structural units and pore morphology of carbon molecular sieve (CMS) membranes. Panel (**a**) shows the single-layer carbon units, highlighting the presence of $sp^3$–C bonds in CMS-600 and the predominantly $sp^2$–C structure in CMS-750. Panel (**b**) presents the three-dimensional CMS models. Panel (**c**) provides a simulated pore morphology, where red indicates the volume occupied by atoms and blue represents the pore volume, illustrating the different pore structures that result from varying pyrolysis conditions [55]. Reproduced from ref. [55] with permission from Elsevier, copyright 2023. (**C**) is a schematic of the gas transport mechanism in CMS/ZSM-5 mixed-matrix membranes (MMMs). The diagram shows various gas molecules ($H_2$, $CO_2$, $O_2$, $N_2$, $CH_4$) moving through the carbon matrix. It highlights the role of the ZSM-5 particles and the interface gap in influencing the selective transport of gas molecules [56]. Reproduced from ref. [56] with permission from Elsevier, copyright 2022

water filtration and high-resolution molecular sieving. Research efforts are focused on mastering the structural control and integration of these materials. Figure 3A provides a valuable schematic comparison between two primary membrane architectures: (a) the pure vertically-aligned CNT (VA-CNTs) membranes and (b) CNT–polymer composite (CNT–CPS) membranes. This figure further details two methods for integrating CNTs into composite structures: (c) depositing a CNT or mixed-matrix layer onto a porous support, and (d) utilizing the CNT layer as a dedicated intermediate layer between a selective polyamide film and the structural support.

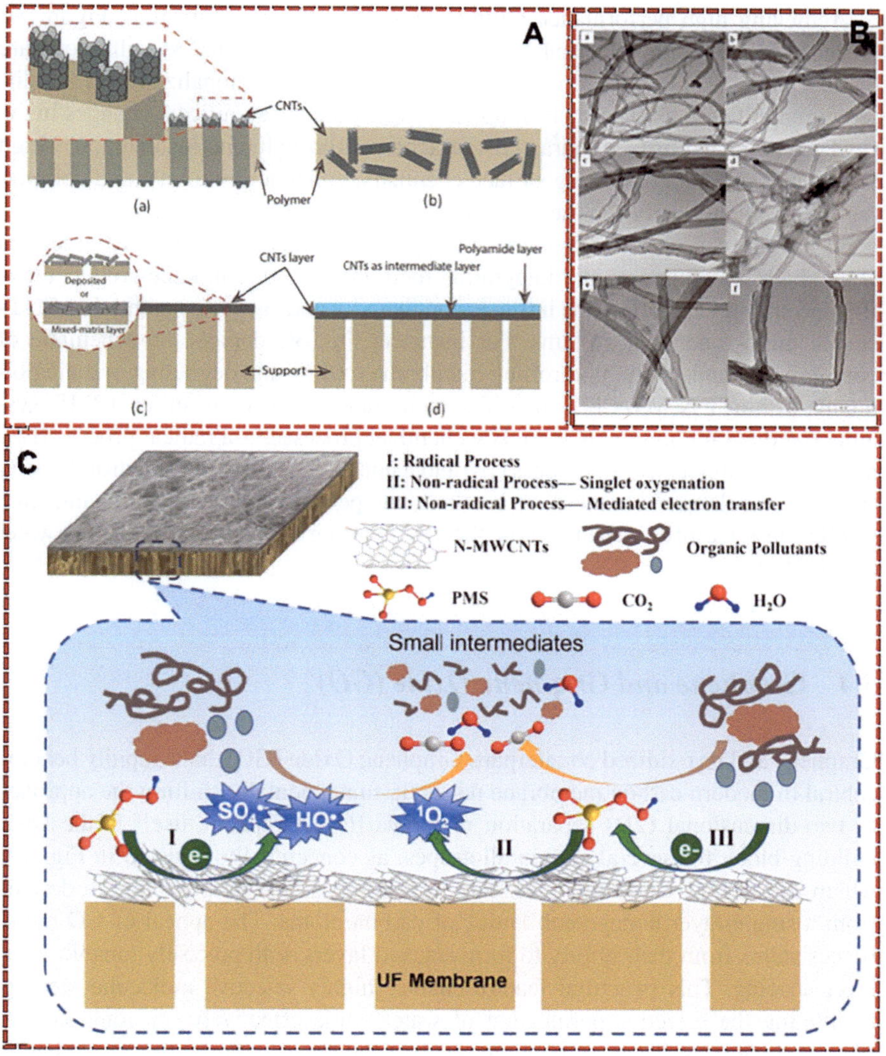


**Fig. 3** (**A**) is a schematic representation that displays the structural differences between (**a**) vertically-aligned CNT (VA-CNTs) membranes and (**b**) CNT-polymer composite (CNT-CPS) membranes. It also illustrates two different methods for incorporating CNTs into CNT-CPS membranes: (**c**) depositing CNTs or a mixed-matrix layer onto a support, and (**d**) using a CNT layer as an intermediate layer between a polyamide layer and a support. Reproduced from ref. [39] with permission from Wiley-VCH GmbH, copyright 2022. (**B**) Shows transmission electron microscope (TEM) images of pristine and functionalized CNTs at different magnifications. Panels (**a**), (**b**), and (**c**) display pristine CNTs, while panels (**d**), (**e**), and (**f**) show functionalized CNTs. The images provide a close-up view of the CNT morphology before and after functionalization [58]. Reproduced from ref. with permission from [58], copyright 2021. (**C**) Is a schematic that shows the possible mechanisms for controlling membrane fouling and degrading organic pollutants in an N-MWCNTs-modified membrane/PMS system. The diagram details three key processes: (I) a radical process, (II) a non-radical singlet oxygen process, and (III) a non-radical mediated electron transfer, all of which work to break down organic pollutants into smaller, less harmful intermediates [59]. Reproduced from ref. [59] with permission from Elsevier, copyright 2022

Achieving high performance often requires chemical modification. Figure 3B presents Transmission Electron Microscope (TEM) images that visually compare the morphology of (a, b, c) pristine CNTs with (d, e, f) functionalized CNTs at different magnifications. These close-up views illustrate the significant changes in the CNT structure and surface characteristics achieved through functionalization, which is a crucial step for controlling surface chemistry, enhancing stability, and optimizing the interaction between the CNT walls and the fluid being transported. Despite these advanced properties, challenges remain in producing defect-free, large-area aligned CNT membranes and integrating them into scalable module architectures. Beyond simple separation, the intrinsic conductivity and catalytic potential of CNTs enable multi-functional systems. For instance, Fig. 3C depicts the schematic of complex mechanisms for controlling membrane fouling and degrading organic pollutants within an N-MWCNTs-modified membrane/Peroxymonosulfate (PMS) system. The diagram highlights three key chemical processes—a radical process (I), a non-radical singlet oxygen process (II), and a non-radical mediated electron transfer (III)—all working in tandem to break down persistent organic pollutants into smaller, less harmful intermediates, thereby merging separation with active water remediation.

### 3.4 Graphene and Graphene Oxide (GO)

Graphene and its oxidized counterpart, Graphene Oxide (GO), have rapidly become central to modern carbon membrane research, fundamentally shifting the approach to two-dimensional (2D) separation materials [60]. Graphene itself is the basic building block for several carbon allotropes, as conceptually depicted in Fig. 4A, which shows how C60 fullerenes, carbon nanotubes, and graphite are all derived from a single-layer honeycomb lattice of carbon atoms. The appeal of GO membranes stems from their ability to form stacked layers with precisely tunable interlayer spacing. This structural feature enables highly selective molecular sieving, facilitating the selective permeation of water while effectively rejecting salts or organic molecules. Their exceptional mechanical strength, flexibility, and inherent electrical conductivity further position GO as a promising material for next-generation nanofiltration and desalination membranes. However, controlling defect density and maintaining uniform layer stacking at industrial scale remains a significant engineering challenge.

Research continues to focus on enhancing and diversifying GO's functionality through chemical modification. Figure 4B schematically details the reaction mechanism of GO with silver chloride (AgCl) and silver (AgNPs) nanoparticles, illustrating how the reaction pathways depend on the pH conditions, specifically involving the transition of phenolic OHgroups to phenolate anions. This functionalization is critical for creating membranes with integrated antimicrobial properties or enhanced surface charge. Beyond liquid separation, GO composites are optimized for gas transport. Figure 4D illustrates the crosslinking sites and the $CO_2$ separation

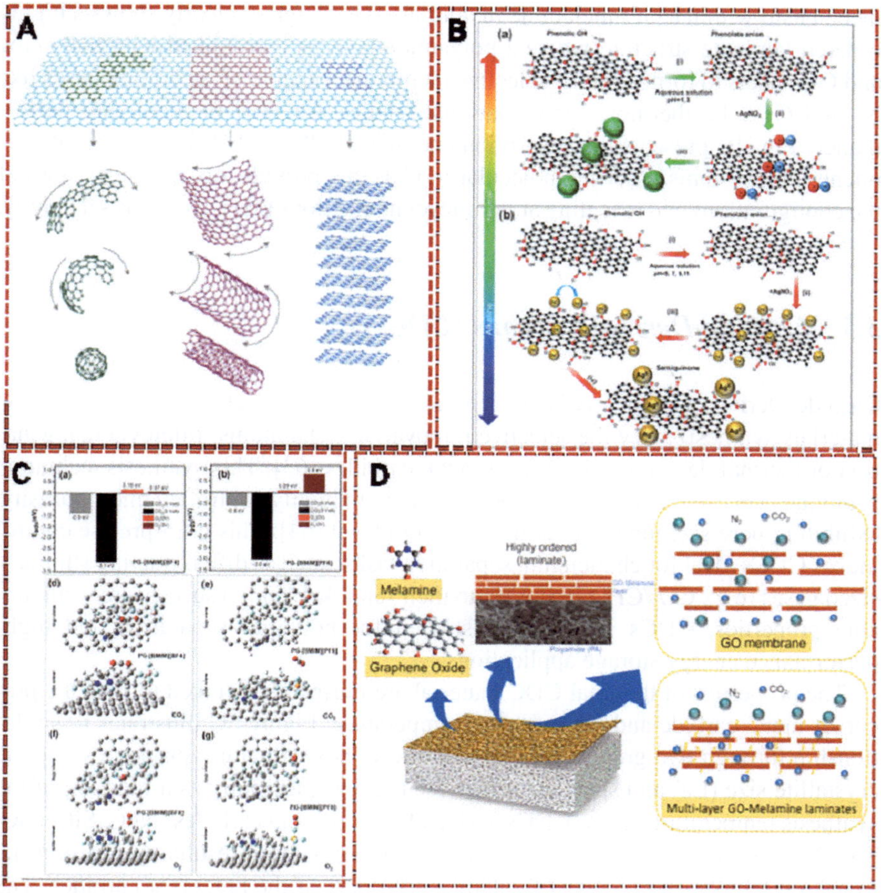


**Fig. 4** (**A**) This diagram shows how C60 fullerene molecules, carbon nanotubes, and graphite can all be conceptualized as being formed from graphene sheets, which are single layers of carbon atoms arranged in a honeycomb lattice [61]. Reproduced from ref. [61] with permission from IEEE Explore, copyright 2009. (**B**) This schematic details the reaction mechanism of graphene oxide (GO) with (**a**) silver chloride nanoparticles (AgCl NPs) and (**b**) silver nanoparticles (AgNPs). It illustrates the different pathways the reactions can take depending on the pH, showing the transition from phenolic OH groups to phenolate anions [62]. Reproduced from ref. [62] with permission from Elsevier, copyright 2022. (**C**) This figure presents the adsorption energies of $CO_2$ and $O_2$ in two systems: (**a**) PG-[BMIM][BF4] and (**b**) PG-[BMIM][PF6]. The accompanying models (**d**–**g**) provide top- and side-views of the molecular configurations, with specific color codes for each atom (Grey: carbon, purple: boron, light blue: fluorine, blue: nitrogen, red: oxygen, and white: hydrogen) [7]. Reproduced from ref. [7] with permission from Elsevier, copyright 2022. (**D**) This diagram illustrates the crosslinking sites and the $CO_2$ separation method for a GO-melamine membrane. It shows the highly ordered laminate structure, where a polyamide (PI) layer is crosslinked with melamine and graphene oxide sheets to form a selective membrane for separating gases like $N_2$ and $CO_2$ [6]. Reproduced from ref. [6] with permission from Elsevier, copyright 2023

mechanism within a GO-melamine membrane. The diagram clearly shows a highly ordered laminate structure where a polyamide layer is crosslinked with melamine and GO sheets to form a robust selective membrane capable of separating gases like $N_2$ and $CO_2$. Furthermore, computational studies are vital in predicting performance: Fig. 4C presents the adsorption energies of $CO_2$ and $O_2$ in two porous graphene (PG) systems, offering molecular models that provide top- and side-views of the configurations and revealing atomic-level interactions that govern gas selectivity.

## 3.5 *Carbide-Derived Carbons (CDCs)*

Carbide-Derived Carbons (CDCs) represent an advanced class of porous carbon materials synthesized by the selective removal of metal atoms (often via halogenation or thermal decomposition) from metal carbides [63]. This synthesis technique is unique because it results in a carbon matrix with highly controlled microporosity, exhibiting pore size tunability at the angstrom level [64]. This ultra-precise control makes CDCs ideal for challenging separation tasks, such as differentiating similarly sized gases (e.g., $CO_2/CH_4$) [65]. Due to their remarkable high surface area and tunable properties, CDCs are actively studied for both gas separation and high-performance energy storage applications.

The properties of the final CDC material are intrinsically linked to both the precursor metal carbide and the synthesis temperature. Figure 5C illustrates this relationship through two graphs: the left plot shows how structural properties like the crystallite size (La) and specific surface area vary significantly as a function of the synthesis temperature across CDCs derived from different carbides (NbC, SiC, TaC, TiC, VC, and ZrC). The right plot further correlates these results by charting the transition temperatures against the lattice constants of the precursor carbides. The resulting microstructures are visually confirmed in Fig. 5D, which presents high-resolution TEM images of CDCs obtained from NbC at different synthesis temperatures (e.g., 1100 °C vs. 1400 °C), revealing the critical role of temperature in microstructural development.

Beyond the fundamental material structure, CDCs are engineered into complex devices. Figure 5A provides a schematic of the fabrication process for a CDC/PA@Ceramic membrane, demonstrating the multi-step approach required: depositing the CDC onto a ceramic support, followed by interfacial polymerization (IP) with TPC and PIP to create a selective polyamide (PA) layer. This integration leverages the structural stability of the CDC layer with the separation capability of the PA layer. Furthermore, the versatile nature of CDCs is highlighted in electrochemical applications: Fig. 5B depicts the use of Transition-Metal- and Nitrogen-Doped Carbide-Derived Carbon/Carbon Nanotube (M-N-CDC/CNT) composites as cathode catalysts for Anion-Exchange Membrane Fuel Cells (AEMFCs). The diagram shows the pyrolysis of CDC and CNT precursors with metal doping agents, with the accompanying graph showcasing the resulting cell voltage and power density curves

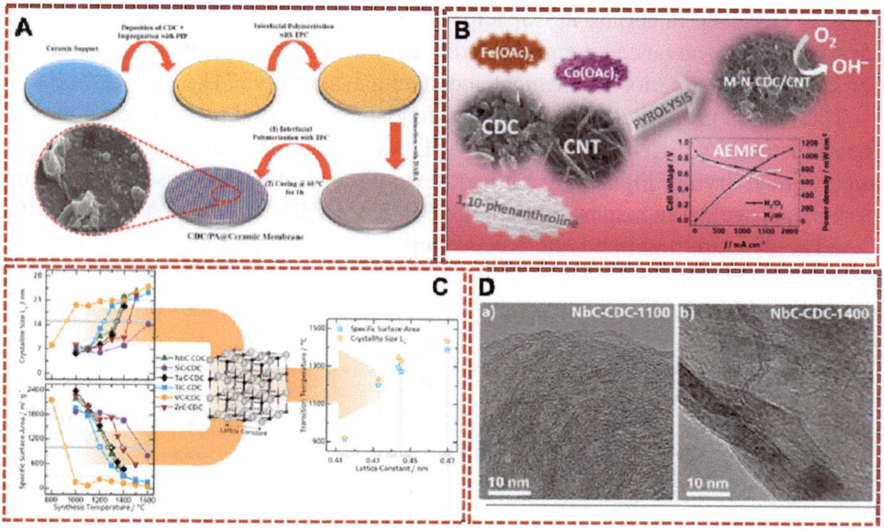


**Fig. 5** (**A**) Provides a schematic representation of the fabrication process for a CDC/PA@Ceramic membrane. The method involves the deposition of CDC onto a ceramic support, followed by interfacial polymerization with TPC and PIP to create a polyamide (PA) layer [66]. Reproduced from ref. [66] with permission from Elsevier, copyright 2022. (**B**) Demonstrates the use of Transition-Metal- and Nitrogen-Doped Carbide-Derived Carbon/Carbon Nanotube (M-N-CDC/CNT) composites as cathode catalysts for Anion-Exchange Membrane Fuel Cells (AEMFCs). The diagram shows the pyrolysis of CDC and CNTs with iron, cobalt, and 1,10-phenanthroline precursors to create the final composite. The accompanying graph shows the cell voltage and power density curves for the AEMFC [67]. Reproduced from ref. [67] with permission from American Chemical Society, copyright 2021. (**C**) Presents two graphs showing the structural properties of various CDCs. The left side shows the crystallite size (La) and specific surface area as a function of synthesis temperature for CDCs derived from NbC, SiC, TaC, TiC, VC, and ZrC. The right plot shows the transition temperatures plotted against the lattice constants of the carbide precursors [68]. Reproduced from ref. [68] with permission from Elsevier, copyright 2021. (**D**) Features high-resolution TEM (HRTEM) images of CDCs obtained from NbC. Panel (**a**) shows the CDC synthesized at 1100 °C, while panel (**b**) shows the material obtained at 1400 °C. The images, taken at 500x magnification, reveal the microstructures of the CDCs at different synthesis temperatures [68]. Reproduced from ref. [68] with permission from Elsevier, copyright 2021

for the AEMFC, demonstrating CDCs' potential in advanced energy conversion and storage.

## 3.6 *Carbon Aerogels*

Carbon Aerogels (CAs) are exceptional porous materials, distinguished by their ultra-low density and highly interconnected nano-networks [69]. Their large surface area and remarkable adjustable pore size make them highly versatile, particularly suitable for capacitive applications and low-pressure gas adsorption [70]. When used as electrode materials in Capacitive Deionization (CDI) systems or as structural supports in advanced membrane architectures, CAs offer enhanced mass transfer kinetics and superior mechanical and chemical stability [71].

The unique properties of aerogels are fundamentally tied to their complex synthesis methods, which often involve sol-gel processing followed by freeze-drying or supercritical drying. Figure 6A outlines several preparation processes for different carbon aerogel composites, including ZIF–8–C, ZIF–8/CMC–CA, and CMC–CA. The diagram highlights the crucial steps, such as mechanical mixing and carbonization, and demonstrates how freeze-drying is used to maintain the delicate three-dimensional hydrogel structure while removing solvent, thus preventing pore

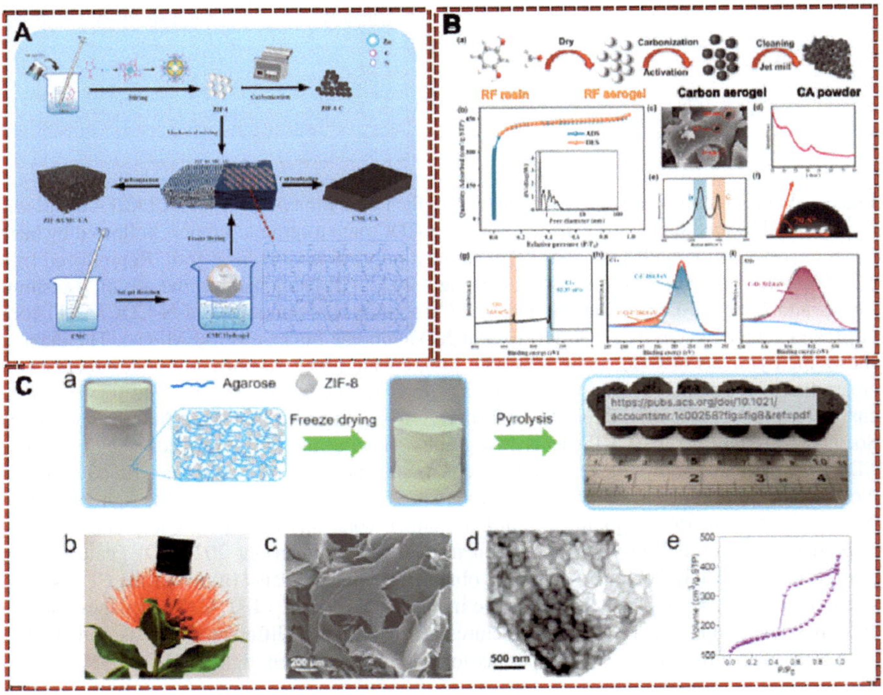


**Fig. 6** (**A**) Outlines the preparation processes for ZIF-8-C, ZIF-8/CMC-CA, and CMC-CA. The diagram shows a mechanical mixing and carbonization process for ZIF-8-C and ZIF-8/CMC-CA. It also depicts a freeze-drying method for creating a CMC hydrogel which is then carbonized to form CMC-CA aeroge l [72]. Reproduced from ref. [72] with permission from Elsevier, copyright 2023. (**B**) Details the preparation procedure and characterization of an industrialized carbon aerogel (CA) powder. Panel (**a**) shows the steps from RF resin to the final CA powder. Panels (**b**) through (**i**) display the results of various analyses, including $N_2$-based physisorption (**b**), SEM imaging (**c**), XRD (**d**), Raman spectra (**e**), contact angle measurement (**f**), and XPS spectra (**g–i**) [73]. Reproduced from ref. [73] with permission from Elsevier, copyright 2022. (**C**) Presents the freeze-drying method for fabricating 3D MOF-derived nanoarchitectures. Panel (**b**) shows a piece of a specific material, ZIF-8/AG-CA-1.0, standing on a flower, demonstrating its lightweight and stable structure. The accompanying images display its morphology via SEM (**c**) and TEM (**d**), while panel (**e**) shows its $N_2$ adsorption-desorption isotherm [74]. Reproduced from ref. [74] with permission from John Wiley & Sons., copyright 2020

collapse to form the final CMC–CA aerogel. The focus on precise structural control is further exemplified by efforts to scale up production. Figure 6B details the preparation procedure for an industrialized carbon aerogel (CA) powder (Panel a). Subsequent panels (b through i) illustrate the suite of characterization techniques required, including $N_2$-based physisorption (b) for surface area analysis, SEM imaging (c) for morphology, and XPS spectra (g,h,i) for chemical composition—all essential for ensuring product consistency.

Advanced research frequently combines the lightweight structure of aerogels with the functionality of Metal-Organic Frameworks (MOFs). Figure 6C presents the freeze-drying method used to fabricate 3D MOF-derived nanoarchitectures. Panel (b) visually demonstrates the extreme low weight and structural integrity of a material, ZIF–8/AG–CA–1.0, by showing it standing delicately on a flower. The accompanying SEM and TEM images (c, d) confirm the interconnected morphology at the nanoscale, while the $N_2$ adsorption-desorption isotherm (e) validates its highly porous structure, underlining the potential of carbon aerogels as highly efficient and tunable materials in both membrane and energy applications.

## *3.7 Emerging Precursors and Green Carbon Sources*

The push for sustainability and cost reduction is fundamentally reshaping the materials landscape for carbon membranes [1]. While polyimides (PIs) currently serve as the performance benchmark for high-selectivity Carbon Molecular Sieve (CMS) membranes due to their robust thermal stability and ability to form intricate ultramicropore structures upon pyrolysis, their adoption is constrained by several factors [75]. These include the high synthesis cost, the complexity of managing toxic, polar aprotic solvents (like NMP or DMF) required for processing, and their overall large environmental footprint.

To address these limitations, there is a rapidly growing interest in emerging precursors derived from biomass and renewable polymers [76]. Materials like cellulose, lignin, and chitosan offer unparalleled advantages in terms of abundance, low cost, and inherent eco-friendliness [77]. For instance, carbon membranes fabricated from cellulose and its derivativeshave demonstrated considerable promise in lab-scale studies [78]. The technical challenge here, however, lies in maintaining structural integrity and high performance during the carbonization process [79, 80]. Since biomass precursors often have lower carbon yields and less rigid molecular structures than PIs, achieving the precise angstrom-level control over pore size distribution—the characteristic trait of high-performance CMSMs—remains an active area of research. Innovations focus on chemical pre-treatment (e.g., cross-linking or additive doping) to improve the precursor's thermal stability and guide the formation of high-selectivity pores [81].

Beyond replacing the precursor polymer itself, research is focused on "greening" the fabrication process for existing high-performing materials [82]. A significant development involves utilizing ionic liquids (ILs) as sustainable alternatives to conventional organic solvents. ILs are generally non-volatile, non-flammable, and can be easily recycled, drastically reducing the environmental impact associated with solvent usage [83]. Using ILs for dissolving and casting PIs or other high-performance polymers offers a tangible path toward cleaner membrane fabrication, potentially simplifying the production cycle and lowering the operational risk without compromising the exceptional performance characteristics inherent to advanced synthetic carbon materials [84]. Furthermore, incorporating sustainable methodologies into the upstream processes, such as utilizing microwave-assisted heating for faster, more energy-efficient pyrolysis, is explored to reduce the overall energy intensity associated with CMS manufacturing [85]. In essence, the future of carbon membrane development involves a dual approach: pursuing true green precursors from biomass, while simultaneously seeking cleaner, greener synthesis routes for established, high-performing synthetic materials.

## 4 Physical and Chemical Characteristics Contributing to Membrane Performance

Carbon-based membranes have gained a lot attention in separation technology because of their exceptional performance characteristics [86]. The physical and chemical properties of these membranes play a key role in determining things like their permeability, selectivity, mechanical strength, and chemical resistance [87]. Understanding how these characteristics work can help researchers design membranes that are more effective and better suited for specific uses, like gas separation [5–10], water treatment [11–14], and other filtration processes.

### *4.1 Physical Characteristics*

#### Pore Size and Distribution

Pore size is arguably the most fundamental physical characteristic of a membrane, as it directly dictates its separation capabilities. The mechanism of separation is fundamentally different depending on whether the membrane is porous or dense.

Porous Membranes (Microfiltration and Ultrafiltration): For microfiltration (MF) and ultrafiltration (UF), separation is a physical sieving process [88]. The membrane acts like a filter, where molecules or particles larger than the pore openings are physically blocked and cannot pass through. To achieve high selectivity and prevent unwanted materials from "leaking" through, these membranes must have a very narrow pore size distribution, meaning the pores are all very similar in size. This ensures consistent and reliable rejection of specific components.

Dense Membranes (Nanofiltration and Reverse Osmosis): In contrast, nanofiltration (NF) and reverse osmosis (RO) membranes are dense and non-porous [89].

Their separation mechanism is not based on sieving but on a solution-diffusion model. In this process, molecules dissolve into the membrane material and then diffuse through it under a pressure gradient. The separation efficiency depends on the difference in the solubility and diffusivity of various molecules within the membrane's polymer matrix. This makes them highly effective at separating even very small species like salts from water.

#### Porosity

Porosity, which is the total void volume within the membrane structure, has a direct and significant impact on the membrane's permeability [90]. A more porous membrane, by definition, has more open space for fluid to flow through, which generally leads to a higher flow rate, or flux, for a given pressure [91]. However, the relationship between porosity and performance is not straightforward [92]. A critical trade-off exists between porosity and selectivity. As porosity increases, the membrane's structural integrity can be compromised, and the pore size distribution may become wider and less uniform [93]. This can lead to a significant loss of selectivity, as it becomes more difficult to consistently reject specific molecules or particles [94]. In essence, a membrane can be very permeable but fail to perform the desired separation if its porosity is too high and poorly controlled.

#### Surface Roughness

Surface roughness is a key factor in determining a membrane's susceptibility to fouling, which is the accumulation of unwanted material on the membrane surface [95]. Fouling can be caused by the deposition of organic matter, biological material, colloids, or inorganic scales [96]. A rougher surface offers a larger effective area for foulants to attach to and provides more "valleys" or crevices where materials can become physically trapped and accumulate [97]. This not only accelerates the fouling process but also makes the membrane more difficult to clean, leading to a faster and more permanent decline in flux and efficiency [98]. Conversely, a smoother membrane surface presents fewer anchoring points for foulants to adhere to, which helps to mitigate fouling and extend the operational life of the membrane [99]. Many membrane manufacturers now focus on creating membranes with low surface roughness to improve their anti-fouling performance [100].

#### Membrane Thickness

The thickness of the active separation layer of a membrane is a crucial variable that affects both flux and durability [87]. According to the solution-diffusion model, flux is inversely proportional to membrane thickness [101]. This means a thinner membrane offers less resistance to the passage of fluid, resulting in a higher flow rate

[102]. This is why researchers constantly strive to make the active layer as thin as possible.

However, this comes with a major trade-off in durability. A very thin membrane can be fragile and more susceptible to physical damage, such as tearing or compaction under high pressure, which can lead to rapid degradation and failure [103]. The development of thin-film composite (TFC) membranes was a significant innovation designed to overcome this very problem. TFC membranes use a thick, porous, and mechanically strong support layer to provide the necessary durability, while a very thin, selective active layer is cast or deposited on top of it [104]. This clever design separates the functions of mechanical support and selective separation, allowing for both high flux and robust performance [105].

## *4.2 Chemical Characteristics*

Membrane performance is a complex interplay of physical and chemical properties [106]. While physical characteristics like pore size and thickness govern the basic mechanics of separation, chemical properties like surface charge and composition dictate the specific interactions between the membrane and the substances being filtered. A membrane's surface charge is a critical chemical characteristic, as it creates electrostatic interactions that either repel or attract charged solutes in the feed water [107]. This is particularly important for NF and RO membranes [108]. The primary separation mechanism is known as Donnan exclusion, where electrostatic repulsion enhances solute rejection [109]. For example, a negatively charged membrane surface can effectively repel negatively charged organic acids, leading to their strong rejection [110]. The magnitude and distribution of this charge can be fine-tuned during the manufacturing process to optimize the membrane for specific applications [111].

Another key property is hydrophilicity/hydrophobicity, which refers to the membrane's affinity for or aversion to water [112]. This has a significant impact on both flux and fouling. Hydrophilic membranes are generally more resistant to fouling by hydrophobic substances [113]. They create a protective hydration layer on the surface that effectively repels hydrophobic foulants, keeping the membrane clean [114]. The surface chemistry can also influence the transport of water and solutes through the membrane, affecting overall flux [115].

The chemical composition of the membrane material, such as cellulose acetate or polyamide, is what dictates all of its inherent chemical and physical properties [116]. The specific functional groups within the polymer chains play a vital role in separation by influencing interactions with different solutes through mechanisms like hydrogen bonding or van der Waals forces [117]. Furthermore, chemical composition determines the membrane's long-term stability and resistance to degradation from chemical agents, such as cleaning solutions like sodium hypochlorite [118].

Finally, for polymeric membranes, the polymer morphology—or the overall structure and arrangement of the polymer chains—is crucial [119]. This morphology affects properties like the effective pore size and the specific transport pathways within the membrane [120]. It includes the shape of internal voids and the packing density of the polymer chains, both of which can be precisely controlled during fabrication to fine-tune the membrane's performance for a given application [121].

# 5 Challenges in the Research and Development of Carbon-Based Membranes

## *5.1 Synthesis Imperfections and Structural Fragility*

The creation of carbon membranes depends on a process called pyrolysis. In this high-temperature process, special polymers are converted into a carbon material with many small pores [122]. A major challenge in this area is making a perfect, defect-free structure [123]. Even small mistakes during the pyrolysis process, like uneven heating or the uncontrolled release of gases, can create tiny cracks or holes in the membrane [124]. These defects act like open pathways, allowing gases to pass through without any separation [125]. This greatly reduces the membrane's ability to separate gases effectively [125, 126]. For example, when separating oxygen from nitrogen, even a single large defect can cause the membrane's selectivity to drop from a very high level to almost zero [127].

This production challenge is made worse by the fragile nature of the final material. The carbon that is produced has a rigid, glassy structure [128]. This means it can break easily when it is being handled or put under mechanical stress [129]. Because of this, the carbon membrane often needs to be placed on a strong support material, like ceramic or metal [130]. However, this creates another problem. When the temperature changes, the carbon layer and its support material expand and contract at different rates. This mismatch can cause the carbon layer to peel away from the support, which creates new and often serious structural problems.

## *5.2 Operational Instability Due to Capillary Condensation*

In addition to problems with their physical structure, carbon membranes have a key limitation in how they operate [39]. This limitation is their sensitivity to a process called capillary condensation. This happens when there are vapors in the gas stream that can turn into a liquid, such as water or hydrocarbons. These vapors stick to the walls of the small pores and eventually turn into a liquid, which blocks the pathways that other, non-condensable gases need to use to pass through the membrane (see Fig. 7).

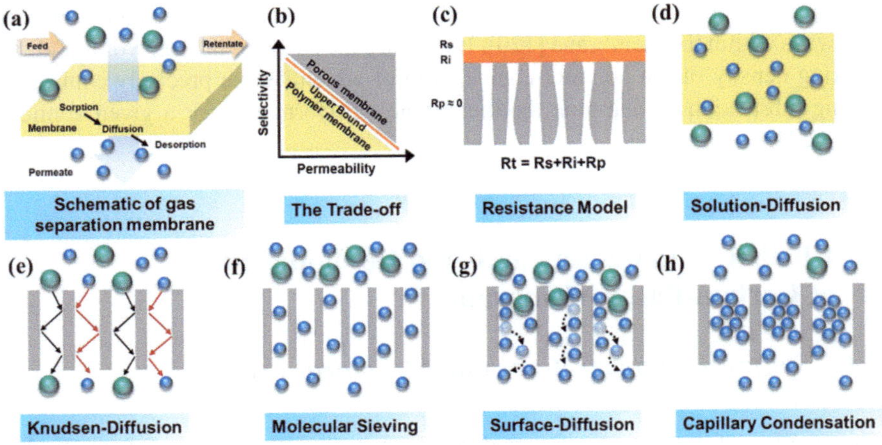


**Fig. 7** (**a**) Schematic representation of gas separation by membrane technology. (**b**) The trade-off between permeability and selectivity. (**c**) Theschematic of a cross-section of the asymmetric composite membrane with corresponding layer resistance. Diffusion mechanisms in polymersand porous materials: (**d**) solution–diffusion, (**e**) Knudsen-diffusion, (**f**) molecular sieving, (**g**) surface diffusion, and (**h**) capillary condensation [9]. Reproduced from ref. [9] with permission from The Royal Society of Chemistry, copyright 2022

This effect is well-known and has been studied extensively. For instance, early experiments by Barrer and Pope showed that when sulfur dioxide was adsorbed by a microporous carbon membrane, it completely blocked the movement of hydrogen gas [131]. This decrease in performance is not a chemical change but a physical blockage [132]. If the condensable vapor is removed, the membrane's performance usually returns to normal [133]. However, in industrial use, the gas streams are rarely pure. Because of this, the reversible but damaging pore blockage makes the technology unsuitable for many applications [85]. This is especially true in environments with moisture or hydrocarbons, where standard polymeric membranes perform much better [134].

## 5.3 *Economic Barriers and Scalability Challenges*

The use of carbon membranes in practical applications is also limited by major economic issues [35]. The production process requires a lot of energy because it needs precise temperature control in an environment without oxygen [135]. This makes the energy costs very high [136]. In addition, the high-performance polymers used to make these membranes, such as polyimides, are much more expensive than the polymers used for standard membranes [137].

It has also been difficult to scale up this complex process for industrial production [138]. To build industrial-sized modules, thousands of kilometers of perfect carbon hollow fibers may be needed [139]. However, many of these fibers break during the pyrolysis process, which leads to low production success rates and a lot of wasted material [140]. Using hybrid designs, where a carbon layer is coated onto a ceramic support, is an alternative, but this approach brings its own difficulties [141]. Defects often form at the boundary between the carbon and the support material [142]. These defects usually need to be sealed after production, which can lower the membrane's overall permeability by 30% to 40%. All of these issues combined make carbon membranes 10 to 100 times more expensive than their polymeric competitors, so they are only used for specialized, high-value purposes [20].

## 5.4 The Competitive Landscape: The Dominance of Polymeric Membranes

In the end, carbon membranes must compete with polymeric technologies, which currently dominate the industrial market [35]. Materials like cellulose acetate or polysulfone provide an attractive mix of flexibility, low cost, and reliable operation [116]. They can handle changes in humidity and pressure much better than brittle carbon membranes, which can be quickly damaged by water causing their pores to collapse [143]. Also, many polymers are more resistant to being damaged by aggressive gases like carbon dioxide [144]. While carbon membranes show better natural selectivity for important gas pairs like $CO_2/CH_4$ or $O_2/N_2$, their brittleness, high cost, and operational instability make them impractical for most commercial uses [52]. This competitive disadvantage is even greater because the manufacturing of polymeric membranes is a very mature technology that has been improved over decades, with scalable and high-yield production methods [145].

## 5.5 Conclusion: A Paradox of Performance and Practicality

To summarize, carbon membranes present a paradox in separation technology [146]. Their very fine porous structure gives them an unmatched ability to perform molecular sieving [147]. However, they remain mostly in the world of academic research because of ongoing and connected challenges. These challenges include their structural brittleness, which limits their use in high-pressure or mechanically stressful environments, and their sensitivity to defects, which requires complex and inefficient production methods. Additionally, their operational instability makes them vulnerable to pore blockage by common condensable gases, and their prohibitive costs prevent them from competing with mature polymeric technologies [148].

### *5.6 Research Directions Toward Practical Solutions*

For carbon membranes to move beyond these limited roles, future research must focus directly on solving these key challenges. Current research efforts are aimed at creating stronger and more stable systems by using hybrid materials and advanced production techniques [35]. For example, combining carbon with metal-organic frameworks (MOFs) can create complex pore structures that resist capillary condensation while still maintaining high selectivity [149]. Similarly, developing advanced production techniques like microwave-assisted pyrolysis shows promise for reducing heat stress during the carbonization step [150]. This could lead to fewer defects (by up to 40%) and improve production efficiency [151]. The great potential of carbon membranes will not be fully realized until major progress is made in defect-free manufacturing, fouling-resistant designs, and scalable production methods. Achieving this progress will demonstrate a successful balance between material innovation and industrial practicality.

## 6 Conclusion and Future Work

This review has traced the development and explained the fundamental properties of carbon-based membranes. These materials, from early carbon sieves to modern graphene structures, offer unmatched theoretical performance in separation technology due to their precisely controlled porous structure and excellent stability in high-temperature and harsh chemical conditions. Their ability to perform molecular sieving continues to set a standard that normal polymers cannot meet.However, this potential has not been widely achieved in an industrial setting. The field is shaped by a central paradox where the special material properties that provide outstanding selectivity are the same properties that cause serious failures in practical use. The most important of these are the material's inherent brittleness, its susceptibility to pore blockage by capillary condensation, and the major difficulties in large-scale, cost-effective manufacturing. These limitations have restricted carbon membranes to very specific applications, preventing them from competing with the flexible and economic benefits of well-established polymeric technologies.

To enable carbon membranes to transition from a laboratory curiosity to a viable industrial technology, a research strategy that directly confronts the challenges identified in this review is essential. Future work must be carefully planned to overcome the intertwined problems of structural fragility, operational instability, and prohibitive cost. A critical area of focus is the inherent mechanical brittleness of these materials. Rather than simply relying on external supports, which can create their own failure points, research should intensify on hybrid architectures and nanocomposite systems. A promising approach involves integrating robust carbon nanomaterials like CNTs or graphene directly into a more flexible polymer matrix to combine the exceptional selectivity of a carbon-based framework with the enhanced durability required for real-world applications. Simultaneously, the persistent problem of capillary condensation needs to be addressed to improve operational stability. One of the most promising avenues for this is pore-surface engineering, which involves chemically modifying the internal surfaces of the pores to be more hydrophilic or amphiphilic. On a larger scale, research into innovative membrane structures, such as hierarchical pore systems formed by combining Metal-Organic Frameworks (MOFs) with carbon, could offer practical solutions for reliably treating the non-ideal gas streams commonly found in industry.

Ultimately, even the most advanced membranes will not see widespread use unless they can be produced economically. Therefore, reducing the high cost of production is non-negotiable and will demand a two-pronged approach cantered on fabrication technology. The development of low-cost, sustainable precursor materials, particularly those derived from abundant biomass sources, is a crucial step to replace the expensive polyimides currently used. In parallel, significant innovation is needed in the pyrolysis process itself. Advanced techniques such as microwave-assisted or laser-induced carbonization could dramatically lower the required energy budget and processing time. These methods also offer the potential for better thermal control, which would minimize defect formation and improve manufacturing yields. By systematically addressing these core issues, the scientific community may finally be able to unlock the immense but unrealized potential of carbon membranes for the next generation of energy-efficient separation technologies.

## References


1. Osman, A.I., et al.: Membrane technology for energy saving: principles, techniques, applications, challenges, and prospects. Adv. Energy Sustain. Res. **5**(5), 2400011 (2024)
2. Li, C., et al.: Carbon-based membrane materials and applications in water and wastewater treatment: a review. Environ. Chem. Lett. **19**(2), 1457–1475 (2021)
3. Kharisov, B.I., Kharissova, O.V.: Carbon Allotropes: Metal-Complex Chemistry, Properties and Applications. Springer (2019)
4. Yang, Z., et al.: Recent advancement of nanostructured carbon for energy applications. Chem. Rev. **115**(11), 5159–5223 (2015)
5. Li, J.-R., Kuppler, R.J., Zhou, H.-C.: Selective gas adsorption and separation in metal–organic frameworks. Chem. Soc. Rev. **38**(5), 1477–1504 (2009)
6. Widakdo, J., et al.: Graphene oxide-melamine nanofilm composite membrane for efficient $CO_2$ gas separation. Sep. Purif. Technol. **323**, 124521 (2023)
7. Widakdo, J., et al.: Bioinspired ionic liquid-graphene based smart membranes with electrical tunable channels for gas separation. Appl. Mater. Today. **27**, 101441 (2022)
8. Widakdo, J., et al.: Mechanism of a self-assembling smart and electrically responsive PVDF–graphene membrane for controlled gas separation. ACS Appl. Mater. Interfaces. **12**(27), 30915–30924 (2020)
9. Widakdo, J., et al.: Switching gas permeation through smart membranes by external stimuli: a review. J. Mater. Chem. A. **10**(32), 16743–16760 (2022)
10. Widakdo, J., et al.: Tailoring of graphene–organic frameworks membrane to enable reversed electrical-switchable permselectivity in $CO_2$ separation. Carbon. **182**, 545–558 (2021)

11. Song, Y., et al.: Porous materials for water purification. Angew. Chem. **135**(11), e202216724 (2023)
12. Austria, H.F.M., et al.: Tailoring the specific crosslinking sites of graphene oxide framework nanosheets for controlled nanofiltration of salts and dyes. J. Clean. Prod. **395**, 136280 (2023)
13. Subrahmanya, T., et al.: An eco-friendly and reusable syringe filter membrane for the efficient removal of dyes from water via low pressure filtration assisted self-assembling of graphene oxide and SBA-15/PDA. J. Clean. Prod. **349**, 131425 (2022)
14. Austria, H.F.M., et al.: Investigation of the pH-mediated fabrication process of pure and polyethyleneimine-crosslinked graphene oxide membranes for desalination and heavy metal ion separation. Carbon. **224**, 119019 (2024)
15. Marchetti, P., Peeva, L., Livingston, A.: The selectivity challenge in organic solvent nanofiltration: membrane and process solutions. Annu. Rev. Chem. Biomol. Eng. **8**(1), 473–497 (2017)
16. Subrahmanya, T., et al.: Effect of functionalized nanodiamonds and surfactants mediation on the nanofiltration performance of polyamide thin-film nanocomposite membranes. Desalination. **555**, 116540 (2023)
17. Widakdo, J., et al.: Positron annihilation spectroscopy for the free volume depth profile analysis of multilayer and 2D materials composite membranes: a review. Sep. Purif. Technol. **322**, 124366 (2023)
18. Abebe, S.H., et al.: Zwitterionic M-PhA modified go nanocomposite membranes for enhanced organic solvent nanofiltration and wastewater treatment with antifouling performance. J. Membr. Sci. **736**, 124596 (2025)
19. Wang, S., et al.: Advances in high permeability polymer-based membrane materials for $CO_2$ separations. Energy Environ. Sci. **9**(6), 1863–1890 (2016)
20. Mollahosseini, A., et al.: Polymeric membranes in carbon capture, utilization, and storage: current trends and future directions in decarbonization of industrial flue gas and climate change mitigation. Energy Environ. Sci. **18**(11), 5025–5092 (2025)
21. Ash, R., Barrer, R.M., Lowson, R.T.: Transport of single gases and of binary gas mixtures in a microporous carbon membrane. J. Chem. Soc., Faraday Trans. **69**, 2166–2178 (1973)
22. Koresh, J., Soffer, A.: Study of molecular sieve carbons. Part 1.—Pore structure, gradual pore opening and mechanism of molecular sieving. J. Chem. Soc., Faraday Trans. **76**, 2457–2471 (1980)
23. Geiszler, V.C., Koros, W.J.: Effects of polyimide pyrolysis conditions on carbon molecular sieve membrane properties. Ind. Eng. Chem. Res. **35**(9), 2999–3003 (1996)
24. Jones, C.W., Koros, W.J.: Carbon molecular sieve gas separation membranes-I. Preparation and characterization based on polyimide precursors. Carbon. **32**(8), 1419–1425 (1994)
25. Iijima, S.: Helical microtubules of graphitic carbon. Nature. **354**(6348), 56–58 (1991)
26. Hinds, B.J., et al.: Aligned multiwalled carbon nanotube membranes. Science. **303**(5654), 62–65 (2004)
27. Holt, J.K., et al.: Fast mass transport through sub-2-nanometer carbon nanotubes. Science. **312**(5776), 1034–1037 (2006)
28. Novoselov, K.S., et al.: Electric field effect in atomically thin carbon films. Science. **306**(5696), 666–669 (2004)
29. Novoselov, K.S., et al.: Two-dimensional atomic crystals. Proc. Natl. Acad. Sci. **102**(30), 10451–10453 (2005)
30. Nair, R., et al.: Unimpeded permeation of water through helium-leak–tight graphene-based membranes. Science. **335**(6067), 442–444 (2012)
31. Liu, M., et al.: Advances in nanofiltration membrane pore size adjustment techniques: a review. Environ. Funct. Mater. (2025)
32. Tan, X., et al.: Application of biochar for the removal of pollutants from aqueous solutions. Chemosphere. **125**, 70–85 (2015)
33. Hou, R., et al.: Current status and advances in membrane technology for carbon capture. Sep. Purif. Technol. **300**, 121863 (2022)

34. Lei, L., et al.: Carbon membranes for $CO_2$ removal: status and perspectives from materials to processes. Chem. Eng. J. **401**, 126084 (2020)
35. Ismail, A.F., David, L.: A review on the latest development of carbon membranes for gas separation. J. Membr. Sci. **193**(1), 1–18 (2001)
36. Sazali, N.: A review of the application of carbon-based membranes to hydrogen separation. J. Mater. Sci. **55**(25), 11052–11070 (2020)
37. Skouteris, G., et al.: The effect of activated carbon addition on membrane bioreactor processes for wastewater treatment and reclamation–a critical review. Bioresour. Technol. **185**, 399–410 (2015)
38. M.B., H.Ä.G.G., Lie, J.A., Lindbråthen, A.: Carbon molecular sieve membranes: a promising alternative for selected industrial applications. Ann. N. Y. Acad. Sci. **984**(1), 329–345 (2003)
39. Barrejón, M., Prato, M.: Carbon nanotube membranes in water treatment applications. Adv. Mater. Interfaces. **9**(1), 2101260 (2022)
40. Liu, G., Jin, W., Xu, N.: Graphene-based membranes. Chem. Soc. Rev. **44**(15), 5016–5030 (2015)
41. Khan, A.U.H.: Water Desalination Using Carbide Derived Carbon (CDC)/Mixed Matrix Membrane. King Fahd University of Petroleum and Minerals (Saudi Arabia) (2016)
42. Moreno-Castilla, C., Maldonado-Hódar, F.: Carbon aerogels for catalysis applications: an overview. Carbon. **43**(3), 455–465 (2005)
43. Davies, A., Yu, A.: Material advancements in supercapacitors: from activated carbon to carbon nanotube and graphene. Can. J. Chem. Eng. **89**(6), 1342–1357 (2011)
44. Kumar, M., Chowdhury, S., Randhawa, J.K.: Emerging trends in membrane-based wastewater treatment: electrospun nanofibers and reticular porous adsorbents as key components. Environ. Sci.:Water Res. Technol. **10**(1), 29–84 (2024)
45. Porada, S., et al.: Review on the science and technology of water desalination by capacitive deionization. Prog. Mater. Sci. **58**(8), 1388–1442 (2013)
46. Tzeng, S.-S., Hung, K.-H., Ko, T.-H.: Growth of carbon nanofibers on activated carbon fiber fabrics. Carbon. **44**(5), 859–865 (2006)
47. Mohan, D., Singh, K.P., Singh, V.K.: Trivalent chromium removal from wastewater using low cost activated carbon derived from agricultural waste material and activated carbon fabric cloth. J. Hazard. Mater. **135**(1–3), 280–295 (2006)
48. Li, H., et al.: Commercially available activated carbon fiber felt enables efficient solar steam generation. ACS Appl. Mater. Interfaces. **10**(11), 9362–9368 (2018)
49. Song, Y.: Carbon molecular sieve membranes for desalination (2014)
50. Zhang, F., et al.: Tailoring nanoporous-engineered sponge fiber molecular sieves with ternary-nested architecture for precise molecular separation. ACS Nano. **15**(8), 13623–13632 (2021)
51. Deng, M., et al.: Next-generation carbon molecule sieve membranes derived from polyimides and polymers of intrinsic microporosity for key energy intensive gas separations and carbon capture. J. Mater. Chem. A. **12**(31), 19806–19838 (2024)
52. Alkandari, S.H.: Advanced and sustainable manufacturing methods of polymeric based membranes for gas separation (2024)
53. Li, L., et al.: A review on the progress in nanoparticle/C hybrid CMS membranes for gas separation. Membranes. **8**(4), 134 (2018)
54. Liu, S., et al.: Carbon molecular sieve membranes derived from hydrogen-bonded organic frameworks for $CO_2/CH_4$ separation. J. Membr. Sci. **678**, 121674 (2023)
55. Xu, R., et al.: High-performance carbon molecular sieve membrane for $C_2H_4/C_2H_6$ separation: molecular insight into the structure-property relationships. Carbon. **201**, 24–36 (2023)
56. Zhang, Y., et al.: Carbon molecular sieve/ZSM-5 mixed matrix membranes with enhanced gas separation performance and the performance recovery of the aging membranes. J. Membr. Sci. **660**, 120869 (2022)
57. Majumder, M., Ajayan, P.: 1.14 Carbon nanotube membranes: a new frontier in membrane science. Compr. Membr. Sci. Eng. **1**(291), e310 (2010)

58. Ngoma, M.M., Mathaba, M., Moothi, K.: Effect of carbon nanotubes loading and pressure on the performance of a polyethersulfone (PES)/carbon nanotubes (CNT) membrane. Sci. Rep. **11**(1), 23805 (2021)
59. Liu, Y., et al.: The nitrogen-doped multi-walled carbon nanotubes modified membrane activated peroxymonosulfate for enhanced degradation of organics and membrane fouling mitigation in natural waters treatment. Water Res. **209**, 117960 (2022)
60. Zhan, M., et al.: Graphene oxide research: current developments and future directions. Nanomaterials. **15**(7), 507 (2025)
61. Xu, C., Li, H., Banerjee, K.: Modeling, analysis, and design of graphene nano-ribbon interconnects. IEEE Trans. Electron. Devices. **56**(8), 1567–1578 (2009)
62. Widakdo, J., et al.: Dual functional GO-Ag incorporated nanocomposite pervaporation membrane with alcohol dehydration performance and enhanced antibacterial property. Mater. Today Chem. **24**, 100985 (2022)
63. Yushin, G., Nikitin, A., Gogotsi, Y.: Carbide derived carbon. Carbon Nanomater., 211–254 (2006)
64. Ania, C.O., Raymundo-Piñero, E.: Nanoporous carbons with tuned porosity. In: Nanoporous Materials for Gas Storage, pp. 91–135. Springer (2019)
65. Singh, J.: Trace Detection in Complex Gas Mixtures by Feedback-Assisted Multipass Raman Scattering. University of South Florida (2025)
66. Waheed, A., et al.: Fabrication of supported carbide-derived-carbon membrane by two phases of interfacial polymerization for oil/water separation. Ceram. Int. **48**(6), 8125–8135 (2022)
67. Lilloja, J., et al.: Transition-metal-and nitrogen-doped carbide-derived carbon/carbon nanotube composites as cathode catalysts for anion-exchange membrane fuel cells. ACS Catal. **11**(4), 1920–1931 (2021)
68. Christians, H., et al.: Mesoporous and crystalline carbide-derived carbons: towards a general correlation on synthesis temperature and precursor structure influence. Carbon. **175**, 215–222 (2021)
69. Zhang, L., et al.: Carbon material-based aerogels for gas adsorption: fabrication, structure design, functional tailoring, and applications. Nanomaterials. **12**(18), 3172 (2022)
70. Sun, M.-H., et al.: Applications of hierarchically structured porous materials from energy storage and conversion, catalysis, photocatalysis, adsorption, separation, and sensing to biomedicine. Chem. Soc. Rev. **45**(12), 3479–3563 (2016)
71. Xu, F., et al.: Selective ion separation by capacitive deionization: a comprehensive review. Materials. **18**(5), 1107 (2025)
72. Wang, Z., et al.: Nanoarchitectonics of heteroatom-doped hierarchical porous carbon derived from carboxymethyl cellulose carbon aerogel and metal-organic framework for capacitive deionization. Int. J. Biol. Macromol. **241**, 124596 (2023)
73. Zhang, X., et al.: Industrially-prepared carbon aerogel for excellent fluoride removal by membrane capacitive deionization from brackish groundwaters. Sep. Purif. Technol. **297**, 121510 (2022)
74. Wang, C., et al.: Large-scale synthesis of MOF-derived superporous carbon aerogels with extraordinary adsorption capacity for organic solvents. Angew. Chem. **132**(5), 2082–2086 (2020)
75. Cai, M., et al.: Remarkably enhanced molecular sieving effect of carbon molecular sieve membrane by enhancing the concentration of thermally rearranged precursors. Sep. Purif. Technol. **341**, 126945 (2024)
76. Isikgor, F.H., Becer, C.R.: Lignocellulosic biomass: a sustainable platform for the production of bio-based chemicals and polymers. Polym. Chem. **6**(25), 4497–4559 (2015)
77. Hussain, S., Akhter, R., Maktedar, S.S.: Advancements in sustainable food packaging: from eco-friendly materials to innovative technologies. Sustainable Food Technol. **2**(5), 1297–1364 (2024)
78. Ahmad, A., et al.: Sustainable cellulose extraction: unlocking the potential of tropical fruit peels for advanced materials. Clean Techn. Environ. Policy. **27**(2), 919–951 (2025)

79. Yang, C., et al.: Advances in the structural engineering and commercialization processes of hard carbon for sodium-ion batteries. J. Mater. Chem. A. **12**(3), 1340–1358 (2024)
80. Wang, Q., Yan, J., Fan, Z.: Carbon materials for high volumetric performance supercapacitors: design, progress, challenges and opportunities. Energy Environ. Sci. **9**(3), 729–762 (2016)
81. He, J., et al.: Liquefiable biomass-derived porous carbons and their applications in $CO_2$ capture and conversion. Green Chem. **24**(9), 3376–3415 (2022)
82. Gilbertson, L.M., et al.: Designing nanomaterials to maximize performance and minimize undesirable implications guided by the Principles of Green Chemistry. Chem. Soc. Rev. **44**(16), 5758–5777 (2015)
83. Pešić, J., et al.: Ionic liquids: review of their current and future industrial applications and their potential environmental impact. Recent Pat. Nanotechnol. **15**(3), 225–244 (2021)
84. Trotta, F.: Exploitation of renewable resources in polymer chemistry. In: Green Chemistry Postgraduate Summer School (online/in-person), p. 50
85. Imtiaz, A., et al.: Challenges, opportunities and future directions of membrane technology for natural gas purification: a critical review. Membranes. **12**(7), 646 (2022)
86. Al-Anzi, B.S., Siang, O.C.: Recent developments of carbon based nanomaterials and membranes for oily wastewater treatment. RSC Adv. **7**(34), 20981–20994 (2017)
87. Nath, K.: Membrane Separation Processes. PHI Learning Pvt. Ltd. (2017)
88. Toyomoto, K., Higuchi, A.: Microfiltration and ultrafiltration. In: Membrane Science and Technology, pp. 301–344. CRC Press (1992)
89. Dach, H.: Comparison of Nanofiltration and Reverse Osmosis Processes for a Selective Desalination of Brackish Water Feeds. Université d'Angers (2008)
90. Manickam, S.S., Gelb, J., McCutcheon, J.R.: Pore structure characterization of asymmetric membranes: non-destructive characterization of porosity and tortuosity. J. Membr. Sci. **454**, 549–554 (2014)
91. Cui, Z., Jiang, Y., Field, R.W.: Fundamentals of pressure-driven membrane separation processes. In: Membrane Technology, pp. 1–18. Elsevier (2010)
92. Lian, C., Zhuge, Y., Beecham, S.: The relationship between porosity and strength for porous concrete. Constr. Build. Mater. **25**(11), 4294–4298 (2011)
93. Sadeghi, I., Kaner, P., Asatekin, A.: Controlling and expanding the selectivity of filtration membranes. Chem. Mater. **30**(21), 7328–7354 (2018)
94. Marchetti, P., et al.: Molecular separation with organic solvent nanofiltration: a critical review. Chem. Rev. **114**(21), 10735–10806 (2014)
95. Koo, C.H., Mohammad, A.W., Talib, M.Z.M.: Review of the effect of selected physicochemical factors on membrane fouling propensity based on fouling indices. Desalination. **287**, 167–177 (2012)
96. Al-Amoudi, A.S.: Factors affecting natural organic matter (NOM) and scaling fouling in NF membranes: a review. Desalination. **259**(1–3), 1–10 (2010)
97. Lejars, M., Margaillan, A., Bressy, C.: Fouling release coatings: a nontoxic alternative to biocidal antifouling coatings. Chem. Rev. **112**(8), 4347–4390 (2012)
98. Shi, X., et al.: Fouling and cleaning of ultrafiltration membranes: a review. J. Water Process Eng. **1**, 121–138 (2014)
99. Zulkefli, N.F., et al.: Recent mitigation strategies on membrane fouling for oily wastewater treatment. Membranes. **12**(1), 26 (2021)
100. Wu, S., et al.: Research progress of marine anti-fouling coatings. Coatings. **14**(9), 1227 (2024)
101. Wijmans, J.G., Baker, R.W.: The solution-diffusion model: a review. J. Membr. Sci. **107**(1–2), 1–21 (1995)
102. Ghidossi, R., Veyret, D., Moulin, P.: Computational fluid dynamics applied to membranes: state of the art and opportunities. Chem. Eng. Process. Process Intensif. **45**(6), 437–454 (2006)
103. Kandagor, V.C.: Nanolayer polymeric coatings to enhance the performance and service life of inorganic membranes for high temperature-high pressure biomass pretreatment and other applications (2014)

104. Saleh, E.A.M., et al.: Recent advances in thin film composite (TFC) membranes development: materials and modification methods. Environ. Sci.: Water Res. Technol. **11**, 1059–1085 (2025)
105. Qiu, L., Sun, Y., Guo, Z.: Designing novel superwetting surfaces for high-efficiency oil–water separation: design principles, opportunities, trends and challenges. J. Mater. Chem. A. **8**(33), 16831–16853 (2020)
106. Seidel, A., Elimelech, M.: Coupling between chemical and physical interactions in natural organic matter (NOM) fouling of nanofiltration membranes: implications for fouling control. J. Membr. Sci. **203**(1–2), 245–255 (2002)
107. Kumar, R., Ismail, A.: Fouling control on microfiltration/ultrafiltration membranes: effects of morphology, hydrophilicity, and charge. J. Appl. Polym. Sci. **132**(21), 42042 (2015)
108. Zhao, Y.-Y., et al.: Role of membrane and compound properties in affecting the rejection of pharmaceuticals by different RO/NF membranes. Front. Environ. Sci. Eng. **11**(6), 20 (2017)
109. Epsztein, R., et al.: Role of ionic charge density in donnan exclusion of monovalent anions by nanofiltration. Environ. Sci. Technol. **52**(7), 4108–4116 (2018)
110. Ng, L.Y., Mohammad, A.W., Ng, C.Y.: A review on nanofiltration membrane fabrication and modification using polyelectrolytes: effective ways to develop membrane selective barriers and rejection capability. Adv. Colloid Interf. Sci. **197**, 85–107 (2013)
111. Shin, M.G., et al.: Positively charged membranes with fine-tuned nanopores for ultrafast and high-precision cation separation. J. Mater. Chem. A. **9**(43), 24355–24364 (2021)
112. Heinz, D., Amado, E., Kressler, J.: Polyphilicity—an extension of the concept of amphiphilicity in polymers. Polymers. **10**(9), 960 (2018)
113. Maximous, N., Nakhla, G., Wan, W.: Comparative assessment of hydrophobic and hydrophilic membrane fouling in wastewater applications. J. Membr. Sci. **339**(1–2), 93–99 (2009)
114. He, K., et al.: Cleaning of oil fouling with water enabled by zwitterionic polyelectrolyte coatings: overcoming the imperative challenge of oil–water separation membranes. ACS Nano. **9**(9), 9188–9198 (2015)
115. McCutcheon, J.R., Elimelech, M.: Influence of membrane support layer hydrophobicity on water flux in osmotically driven membrane processes. J. Membr. Sci. **318**(1–2), 458–466 (2008)
116. Islam, M.D., et al.: Cellulose acetate-based membrane for wastewater treatment—A state-of-the-art review. Mater. Adv. **4**(18), 4054–4102 (2023)
117. Wiśniewska, M.: A review of temperature influence on adsorption mechanism and conformation of water soluble polymers on the solid surface. J. Dispers. Sci. Technol. **32**(11), 1605–1623 (2011)
118. Mulyati, S., et al.: Investigation on the stability and durability of PVDF membranes under prolonged exposure to household bleach cleaning agents. S. Afr. J. Chem. Eng. **49**(1), 285–294 (2024)
119. Ren, J., Wang, R.: Preparation of polymeric membranes. In: Membrane and Desalination Technologies, pp. 47–100. Springer (2010)
120. Kong, C.S., et al.: Influence of pore-size distribution of diffusion layer on mass-transport problems of proton exchange membrane fuel cells. J. Power Sources. **108**(1–2), 185–191 (2002)
121. Astorino, C., et al.: Advancements in gas separation for energy applications: exploring the potential of polymer membranes with intrinsic microporosity (PIM). Membranes. **13**(12), 903 (2023)
122. Salleh, W.N.W., et al.: Precursor selection and process conditions in the preparation of carbon membrane for gas separation: a review. Sep. Purif. Rev. **40**(4), 261–311 (2011)
123. Chen, H., et al.: A defect-free principle for advanced graphene cathode of aluminum-ion battery. Adv. Mater. **29**(12), 1605958 (2017)
124. Oloruntoba, A., Zhang, Y., Hsu, C.S.: State-of-the-art review of fluid catalytic cracking (FCC) catalyst regeneration intensification technologies. Energies. **15**(6), 2061 (2022)
125. Pandey, P., Chauhan, R.: Membranes for gas separation. Prog. Polym. Sci. **26**(6), 853–893 (2001)
126. Bernardo, P., Drioli, E., Golemme, G.: Membrane gas separation: a review/state of the art. Ind. Eng. Chem. Res. **48**(10), 4638–4663 (2009)

127. Park, H.B., et al.: Maximizing the right stuff: the trade-off between membrane permeability and selectivity. Science. **356**(6343), eaab0530 (2017)
128. Menéndez-Díaz, J., Martín-Gullón, I.: Types of carbon adsorbents and their production. In: Interface Science and Technology, pp. 1–47. Elsevier (2006)
129. Broek, D.: The Practical Use of Fracture Mechanics. Springer (2012)
130. Wang, Y.-J., Wilkinson, D.P., Zhang, J.: Noncarbon support materials for polymer electrolyte membrane fuel cell electrocatalysts. Chem. Rev. **111**(12), 7625–7651 (2011)
131. Uhlhorn, R., Burggraaf, A.: Gas separations with inorganic membranes. In: Inorganic Membranes Synthesis, Characteristics and Applications, pp. 155–176. Springer (1991)
132. Polyakov, Y.S., Zydney, A.L.: Ultrafiltration membrane performance: effects of pore blockage/constriction. J. Membr. Sci. **434**, 106–120 (2013)
133. Nitsche, V., Ohlrogge, K., Stürken, K.: Separation of organic vapors by means of membranes. Chem. Eng. Technol. **21**(12), 925–935 (1998)
134. Buonomenna, M.G.: Membrane processes for a sustainable industrial growth. RSC Adv. **3**(17), 5694–5740 (2013)
135. Zhao, Q., et al.: Spinels: controlled preparation, oxygen reduction/evolution reaction application, and beyond. Chem. Rev. **117**(15), 10121–10211 (2017)
136. Council, N.R., et al.: Hidden Costs of Energy: Unpriced Consequences of Energy Production and Use. National Academies Press (2010)
137. Tan, X., Rodrigue, D.: A review on porous polymeric membrane preparation. Part II: production techniques with polyethylene, polydimethylsiloxane, polypropylene, polyimide, and polytetrafluoroethylene. Polymers. **11**(8), 1310 (2019)
138. Bell, T.A.: Challenges in the scale-up of particulate processes—an industrial perspective. Powder Technol. **150**(2), 60–71 (2005)
139. Stucki, M.T.: Internal Pore Functionalization for Applied Polymer Membranes. ETH Zurich (2017)
140. Zaman, C.Z., et al.: Pyrolysis: a sustainable way to generate energy. In: Pyrolysis, p. 1 (2017)
141. Shearer, C.J., Cherevan, A., Eder, D.: Application and future challenges of functional nanocarbon hybrids. Adv. Mater. **26**(15), 2295–2318 (2014)
142. Gerber, I.C., Serp, P.: A theory/experience description of support effects in carbon-supported catalysts. Chem. Rev. **120**(2), 1250–1349 (2019)
143. Gittleman, C.S., Coms, F.D., Lai, Y.-H.: Membrane durability: physical and chemical degradation. Polym. Electrolyte Fuel Cell Degrad., 15 (2011)
144. Scott, G.: Polymers and the Environment. Royal Society of Chemistry (1999)
145. Figoli, A., Cassano, A., Basile, A.: Membrane Technologies for Biorefining. Woodhead Publishing (2016)
146. Belfort, G.: Membrane filtration with liquids: a global approach with prior successes, new developments and unresolved challenges. Angew. Chem. **131**(7), 1908–1918 (2019)
147. Chen, T., et al.: Elucidating the impact of porous organic cage on thin film nanocomposite membranes for elevated nanofiltration. Desalination. **576**, 117362 (2024)
148. Siegelman, R.L., Kim, E.J., Long, J.R.: Porous materials for carbon dioxide separations. Nat. Mater. **20**(8), 1060–1072 (2021)
149. Lin, H., et al.: Metal–organic frameworks for water harvesting and concurrent carbon capture: a review for hygroscopic materials. Adv. Mater. **36**(12), 2209073 (2024)
150. Alam, S., Khan, A.: Microwave-assisted pyrolysis for waste plastic recycling: a review on critical parameters, benefits, challenges, and scalability perspectives. Int. J. Environ. Sci. Technol. **21**(5), 5311–5330 (2024)
151. Gunasekaran, A., et al.: Improving productivity and quality in manufacturing organizations. Int. J. Prod. Econ. **36**(2), 169–183 (1994)

# Innovative Synthesis Approaches for High-Performance Carbon-Based Membranes

**Muhammad Arrayyan Pasha Marseno, Abraham Nathan Sunanta, Aghniya Erytama Ramadhan, Januar Widakdo, and Isnaini Rahmawati**

## 1 Introduction

Carbon-Based Membranes has grown increasingly important in developing functional membranes, especially for environmental remediation and energy applications [1]. This chapter discussed some synthesis techniques—Chemical Vapor Deposition (CVD), Pyrolysis, Sol-Gel Processing, etc.—with focused on their unique mechanisms and resulting carbon-based membranes [2]. While all these methods produce carbon-rich structures, they are different substantially in starting materials, processing environments, and heat controls [3]. CVD, for instance, relies on reactions between gaseous compounds over heated substrates, which creates even carbon films where structural parameters can be tightly controlled [4]. Pyrolysis works differently, it breaks down organic matter in oxygen-deprived environments to produce either amorphous carbon or partially graphitized structures, depending on conditions [5]. The sol-gel approach merges solution chemistry with heat processing to generate either carbon-ceramic composite frameworks or materials with controlled porosity [6]. Through careful analysis and comparison, this topic emphasizes the significance of these synthesis methods in customizing carbon materials for specific applications, demonstrating their continuing relevance in the future.

M. A. P. Marseno · A. N. Sunanta · A. E. Ramadhan · J. Widakdo
Department of Physics, Faculty of Mathematics and Natural Sciences, Universitas Indonesia, Depok, Indonesia

I. Rahmawati (✉)
Department of Chemistry, Faculty of Mathematics and Natural Sciences, Universitas Indonesia, Depok, Indonesia
e-mail: isnaini.rahmawati@sci.ui.ac.id



H. Abdullah, J. Widakdo (eds.), *Membrane Physics of Carbon-Based Materials*, Engineering Materials, https://doi.org/10.1007/978-981-95-5605-2_2

## 2 Synthesis Methods (Chemical Vapor Deposition, Pyrolysis, Sol-Gel, Etc.)

Carbon-based membranes are synthesized using various methods that influence their structure, porosity, and separation performance [7]. Common synthesis techniques include carbonization of polymeric precursors [8], chemical vapor deposition (CVD) [9], pyrolysis [10], and the sol-gel process [11]. In the carbonization method, organic polymers such as polyimides or phenolic resins are thermally treated under an inert atmosphere, resulting in a rigid carbon matrix with tunable pore sizes [12]. CVD involves the deposition of carbon onto a porous support from hydrocarbon gases at high temperatures, forming ultra-thin, selective carbon layers [13]. Pyrolysis similarly involves the thermal decomposition of organic compounds but allows more control over carbon morphology depending on the precursor and conditions used [14].

The sol-gel method, although less common for carbon membranes, enables the formation of hybrid organic-inorganic materials that can be carbonized to create porous carbon structures [15]. Each of these techniques can be tailored by adjusting parameters like temperature, heating rate, and precursor type to achieve desired membrane characteristics such as pore size distribution, surface area, and mechanical stability [16]. The choice of synthesis method directly affects the membrane's separation efficiency, thermal resistance, and applicability in gas separation, water purification, or organic solvent nanofiltration [17]. Researchers continue to refine these techniques to produce membranes with higher selectivity [18], permeability [19], and structural integrity for industrial-scale applications [20].

### *2.1 Chemical Vapor Deposition (CVD)*

Chemical Vapor Deposition (CVD) is a cornerstone technique in materials science, vital for fabricating high-quality thin films and coatings [21]. Known for its precision and versatility, CVD has become critical in synthesizing materials where purity, uniformity, and performance are paramount [22]. The process centers on the use of vapor-phase chemicals—known as precursors—which undergo chemical reactions or decomposition when exposed to an elevated temperature near, or at, the surface of a solid substrate [23]. Figure 1 provides a comprehensive diagram illustrating the full CVD process flow.

The process begins with the rigorous cleaning and preparation of the substrate. Materials such as silicon, copper, or ceramics are commonly chosen for their thermal stability and compatibility with various thin films [24]. Effective surface preparation is essential to remove contaminants, enhance film adhesion, and promote uniform nucleation during deposition [25]. Next, volatile precursor gases are introduced into the reaction chamber. These gases—often hydrocarbons (like $CH_4$) or organometallics—are transported by a carrier gas, typically argon or nitrogen [25].

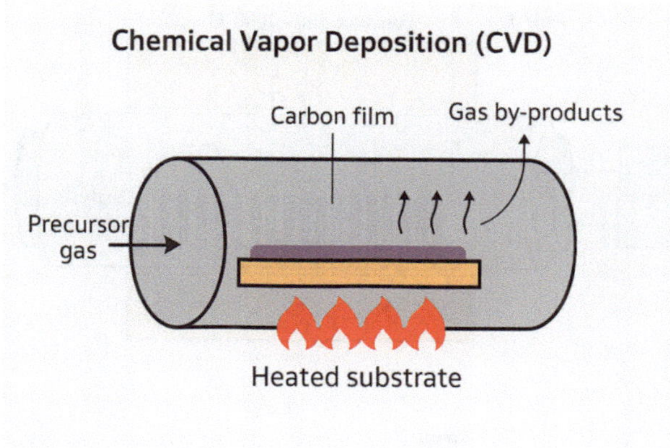


**Fig. 1** Diagram process of CVD

The flow rates and ratios of these gases are meticulously regulated to control film growth and ensure homogeneous gas-phase distribution within the chamber [26]. The substrate is then heated to a specific temperature, often ranging from 400 °C to 1100 °C depending on the desired film material [27]. At this elevated temperature, the precursors begin to react either in the gas phase or upon contacting the heated substrate surface, initiating the deposition process. Figure 2a provides a schematic example of a typical CVD process for graphene synthesis, using $CH_4$ gas as the precursor.

Chemical reactions—which may involve thermal decomposition or surface-catalyzed processes—occur as the gaseous precursors decompose or react on the hot substrate, resulting in the accumulation of a thin, solid film [28]. Figure 2b shows a photograph of a copper (Cu) foil placed on a heating stage before being loaded into a rapid thermal processing reactor, illustrating the type of equipment used in this high-temperature process. The final film's quality, morphology, and crystalline structure depend critically on factors such as temperature, pressure, gas composition, and flow dynamics. During the reaction, gaseous byproducts (e.g., hydrogen, carbon monoxide, or other volatile compounds) are formed and continuously purged from the chamber by the carrier gas, maintaining the necessary reaction environment and preventing contamination of the growing film. What makes CVD particularly valuable is its ability to fine-tune film thickness, composition, and microstructure, making it suitable for both research and industrial-scale production. The applications of CVD are vast: in semiconductors, it deposits layers like silicon dioxide for integrated circuits; in nanotechnology, it enables the controlled growth of carbon nanotubes and graphene; and in engineering, it is invaluable for producing protective coatings, membrane materials, and ceramic composites where strength in extreme environments is essential.

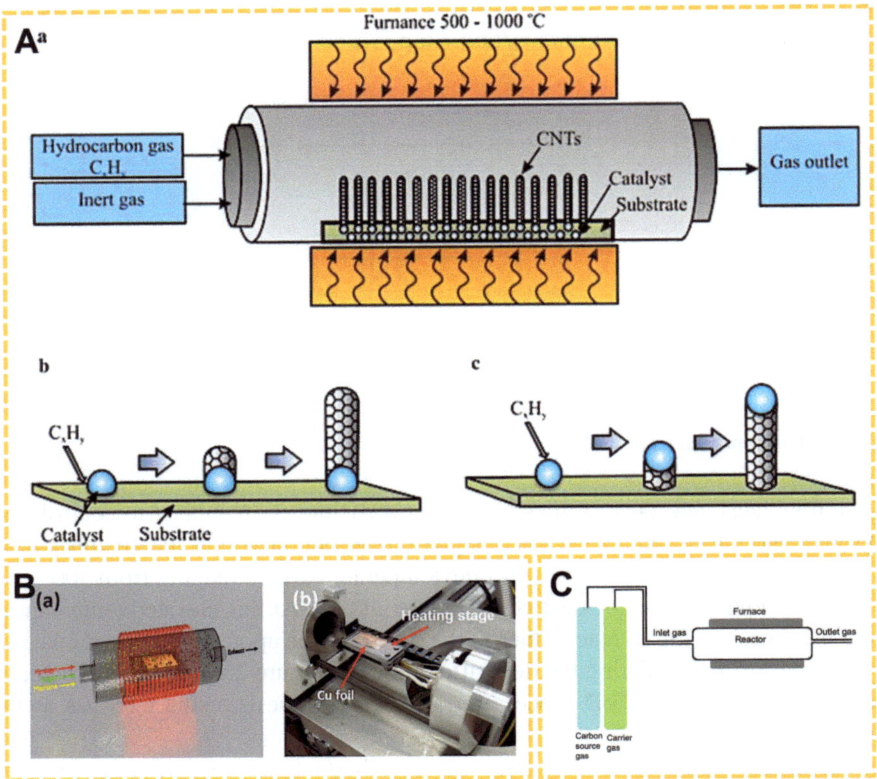


**Fig. 2** (**A**) Schematic representation of the CVD process. A Simplified scheme of (**a**) CVD reactor for CNTs synthesis; (**b**) base-growth model of CNTs growth mechanism; (**c**) tip-growth model of CNTs growth mechanism [29]. Reproduced from ref. [29] with permission from Elsevier, copyright 2023. (**B**) Schematic of typical CVD of graphene using $CH_4$ gas as a precursor (**a**). Photograph of a Cu foil kept on a heating stage before loading in a rapid thermal processing cold wall reactor (**b**). Reproduced from ref. [29] with permission from Elsevier, copyright 2006. (**C**) Schematic of the typical CVD reactor [30]. Reproduced from ref. [30] with permission from Elsevier, copyright 2019

## 2.2 *Pyrolysis*

Pyrolysis is a thermochemical process used to convert organic precursors into carbon-rich materials [31]. Operating in an oxygen-free environment, this method allows controlled decomposition, making it ideal for producing amorphous or graphitic carbon structures with tunable properties [31]. The pyrolysis process begins with the selection of an appropriate organic precursor, such as a polymer, resin, or biomass-derived compound [32]. The choice of precursor greatly influences the resulting carbon structure, and it is typically shaped or cast into the desired form before thermal treatment [16]. To prevent oxidation or combustion, the precursor is placed in a tube furnace and exposed to a flow of inert gas such as nitrogen or argon

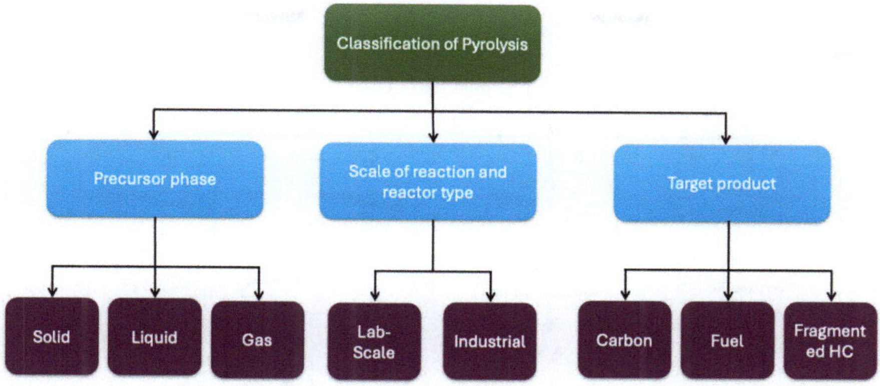


**Fig. 3** Different classification pathways of pyrolysis process [40]. Reproduced from ref. [40] with permission from Oxford University Press, copyright 2021

[33]. This protective atmosphere ensures that carbonization occurs in a controlled, oxygen-free environment, which is crucial for producing stable carbon materials [34]. The sample is gradually heated at a controlled rate, usually between 2 °C to 10 °C per minute, to the target pyrolysis temperature, typically in the range of 600 °C–1000 °C [35]. During this stage, the precursor undergoes thermal decomposition, releasing volatiles such as water vapor, carbon dioxide, and small hydrocarbons [36].

Once the desired temperature is reached, it is held constant for a designated period—ranging from several minutes to hours—to allow complete carbonization [34]. The organic structure undergoes molecular rearrangement, forming a carbon-rich matrix whose properties depend on the duration and temperature of the pyrolysis [37]. After carbonization, the system is allowed to cool under a continued inert gas flow to avoid reoxidation [38]. The resulting carbon material may then undergo additional treatments such as activation, surface functionalization, or graphitization to enhance porosity, conductivity, or other application-specific properties [39].

The pyrolysis process can be categorized based on three major aspects: the precursor phase, the scale of reaction and reactor type, and the target product [40]. As shown in Fig. 3, the precursor phase determines whether the starting material is in a solid, liquid, or gaseous state, which directly affects heat transfer and decomposition behavior. The second classification considers the reaction scale and reactor configuration, ranging from lab-scale systems used for mechanistic or fundamental studies to industrial setups designed for large-scale carbon or fuel production. The third classification focuses on the target product, which may include carbon materials, fuel, or fragmented hydrocarbons, depending on the process objectives and operating parameters. This categorization provides a structured understanding of how process design and material selection influence the outcome and application of pyrolysis-derived products.

Pyrolysis has been extensively utilized in diverse technological and industrial applications, as illustrated in Fig. 4a, the process finds four major areas of

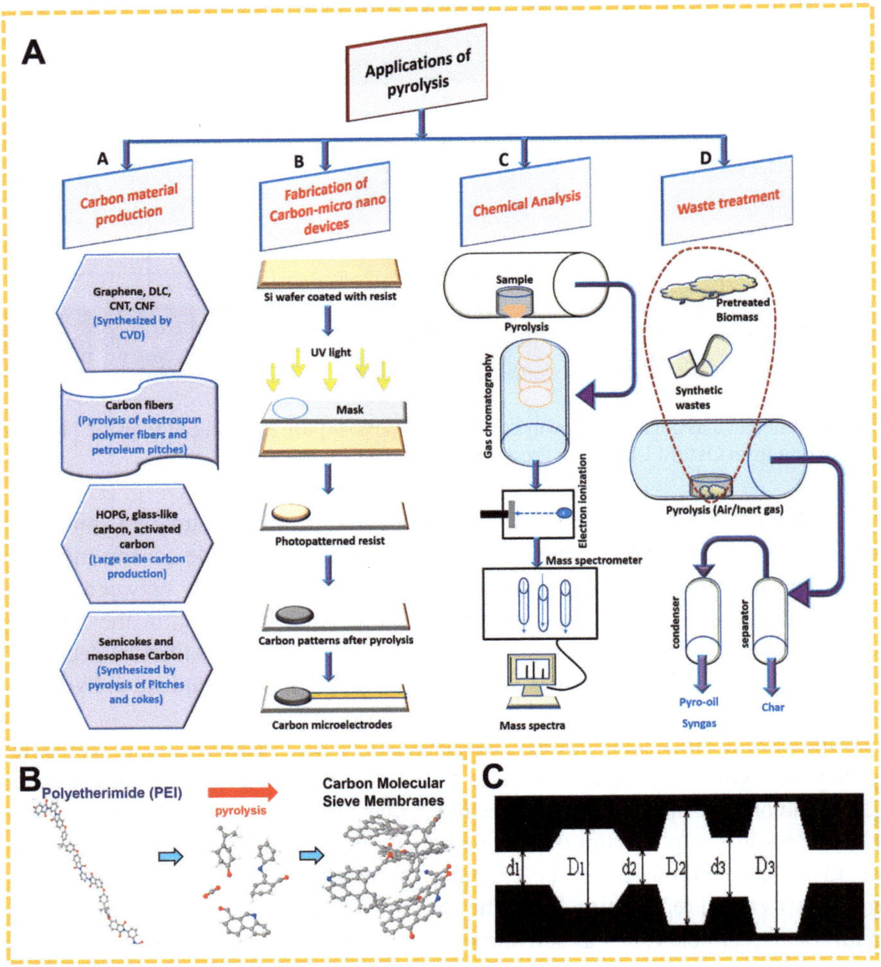


**Fig. 4** (**A**) schematic representation of classification of applications of pyrolysis into four major areas: (**a**) carbon material production, (**b**) fabrication of carbon-micro nano devices, (**c**) chemical analysis of unknown samples by Py-GC–MS, (**d**) treatment of waste. CNF, carbon nanofibers; HOPG, highly oriented pyrolytic graphite [40]. Reproduced from ref. [40] with permission from Oxford University Press, copyright 2021. (**B**) computational analysis of carbon molecular sieve membrane formation upon polyetherimide pyrolysis [41]. Reproduced from ref. [41] with permission from Elsevier, copyright 2017. (**C**) Idealized structure of a pore in a carbon material [42]

implementation: (a) carbon material production, where various carbon allotropes such as graphene, carbon nanotubes, and carbon fibers are synthesized; (b) fabrication of carbon micro- and nanodevices, involving the photopatterning of polymer films followed by pyrolysis to generate conductive carbon microelectrodes; (c) chemical analysis, where Py-GC–MS is applied to characterize the decomposition products of complex samples; and (d) waste treatment, in which synthetic and biomass wastes are thermally converted into syngas, pyro-oil, and char. Additionally, computational studies such as those involving polyetherimide (PEI) pyrolysis provide molecular insights into the formation of carbon molecular sieve membranes (Fig. 4b) [41]. Figure 4c shows an idealized structure of a pore in a carbon material illustrates how pores are interconnected through constrictions and wider cavities, forming a complex network that controls gas permeability, adsorption, and molecular sieving properties. The alternating narrow (d) and wide (D) regions represent the pore throats and pore bodies, respectively, which determine the diffusion pathways and selective transport behavior of gases or liquids through the carbon matrix [42]. Understanding this pore geometry is essential for designing carbon-based materials for separation membranes, catalysis, and energy storage applications.

## *2.3 Sol–Gel*

The sol–gel method is a wet chemical approach for synthesizing ceramics and carbon-ceramic composites [43]. Known for its versatility and molecular-level control, sol-gel processing is ideal for forming homogeneous, porous materials with fine-tuned chemical and physical characteristics [44]. The process starts by preparing a homogeneous solution or sol, which typically involves mixing metal alkoxides or metal salts with water and an alcohol solvent [45]. Hydrolysis and condensation reactions, often catalyzed by acids or bases, lead to the formation of a colloidal suspension composed of fine particles or polymeric chains [46]. As the sol evolves, it undergoes gelation—a process in which the suspended particles or polymers link together, forming a three-dimensional gel network [47]. This network traps the solvent within its structure, resulting in a wet, semi-solid gel. The gelation time and structure depend on precursor concentration, pH, and temperature [48].

After gelation, the gel is dried to remove the trapped solvent. This can be done through ambient drying, vacuum drying, or supercritical drying depending on the desired properties [49]. During this phase, the gel often shrinks and becomes more porous, forming a dry, solid xerogel. The dried gel is then subjected to thermal treatment [50]. If performed in air (calcination), this step decomposes organic residues and converts the gel into an oxide ceramic [51]. Alternatively, heating under inert conditions (carbonization) transforms the gel into a carbon-ceramic hybrid with enhanced conductivity and thermal stability [52]. The final material may be functionalized to tailor its properties. Additives such as metal nanoparticles, polymers, or carbon nanotubes can be incorporated to improve catalytic activity, mechanical strength, or selectivity, depending on the end-use application [53].

Figure 5 illustrates the formation and morphology of a sol–gel–derived carbon nanostructured material. In panel (a), the schematic diagram shows the synthesis pathway of **SMIP@PCNS-gel**, where the sol–gel process enables the integration of molecularly imprinted polymers (SMIP) within a porous carbon nanosphere (PCNS) network, forming a hybrid gel structure. Panels (b) and (c) display the SEM images of the pure PCNS-gel, which exhibits a uniform and interconnected porous structure, while panels (d) and (e) show the **SMIP@PCNS-gel**, where the surface

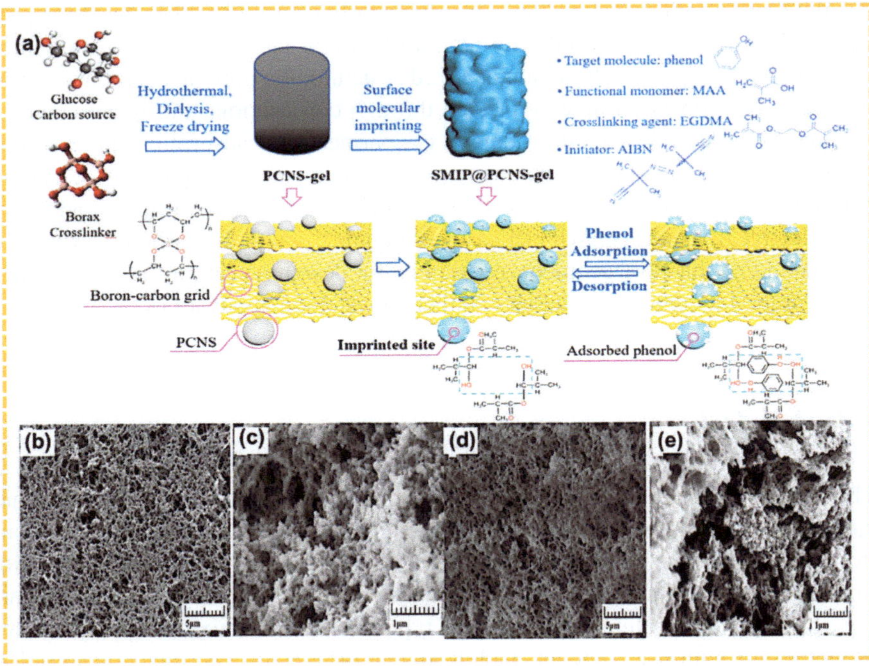


**Fig. 5** (**a**) Graphic illustration showing the synthesis of SMIP@PCNS-gel. SEM images of (**b**, **c**) PCNS-gel and (**d**, **e**) SMIP@PCNS-gel at several intensities [54]. Reproduced from ref. [54] with permission from Elsevier, copyright 2022

becomes rougher and denser due to the incorporation of the imprinted polymer. These morphological differences confirm the successful formation of the composite through the sol–gel process [54].

## 2.4 *Electrospinning and Melt Extrusion*

Fiber-based carbon materials can also be produced through spinning techniques, such as electrospinning and melt extrusion, which allow the fabrication of continuous polymeric precursors that can later be carbonized. Electrospinning employs a high voltage to create a jet from a liquid solution or melt, resulting in a nanofibrous structure [55], meanwhile, melt extrusion shapes polymer melts under elevated temperature and pressure to produce membranes. Compared to electrospinning, which generates membranes with extremely fine pores and large surface areas, melt extrusion is generally more sustainable and easier to scale up since it does not require solvents [56]. Figure 6 illustrates the different production routes of spun carbon fibers (CF) from different types of precursors. Electrospinning utilizes an electric field to draw fine fibers from a polymer solution, such as polyvinyl alcohol (PVA),

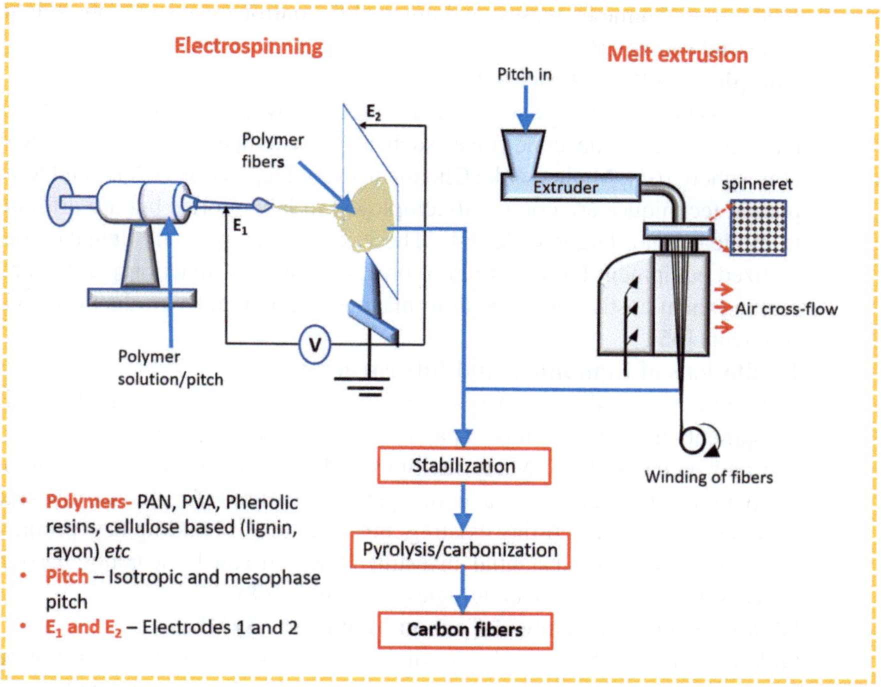


**Fig. 6** Production of spun CF by different routes using different precursors. *PVA* polyvinyl alcohol [40]. Reproduced from ref. [40] with permission from Oxford University Press, copyright 2021

while melt extrusion involves the mechanical drawing of polymer melts through a spinneret to form continuous filaments. Both methods generate polymeric fibers that undergo stabilization and carbonization steps to yield conductive and mechanically stable carbon fibers suitable for various functional applications [40].

1. **Challenges in Large-Scale Production of Advanced Materials**

   The transition from laboratory-scale experiments to large-scale production presents a significant challenge in the development of advanced materials [57]. Although many materials exhibit exceptional properties when synthesized in small quantities, reproducing these properties at industrial scales often proves difficult [58]. Three major challenges—material consistency, complex synthesis processes, and limitations in equipment and infrastructure—pose barriers to efficient mass production [59].

   (a) **Material Consistency**

   One of the foremost challenges in scaling up production is ensuring material consistency. In the case of carbon-based materials, the properties of the final product are highly sensitive to the type and quality of carbon precursors used [60]. Variations in raw materials can lead to inconsistencies in structure, porosity, or conductivity, which in turn affect the material's performance [61]. This lack of uniformity becomes more pronounced as production volume increases, making quality control a persistent obstacle in industrial settings [62].

(b) **Complex Synthesis Processes**

The synthesis of advanced materials often involves complex processes that require extreme conditions, such as high temperatures and an inert atmosphere [63]. Methods like Chemical Vapor Deposition (CVD) and templating techniques are commonly employed in laboratories but are difficult to replicate on a larger scale [64]. These techniques not only demand specialized equipment but also require precise control of reaction parameters, making them costly and difficult to manage in continuous production environments [65].

(c) **Limitations of Equipment and Infrastructure**

Even if the synthesis process is well understood, scaling it up demands equipment and infrastructure capable of maintaining the same level of control and precision [66]. Many facilities lack the advanced reactors, high-temperature furnaces, or vacuum systems required for industrial-scale production [67]. Retrofitting existing infrastructure or building new production lines involves substantial investment, which can be a major barrier, especially for emerging technologies or startups [68].

2. **Cost Analysis and Efficiency of Carbon Membrane Synthesis**

While carbon membranes hold promise for various separation and filtration applications, their widespread adoption is often hindered by cost-related issues [68]. A thorough understanding of the key cost drivers, potential cost reduction strategies, and the trade-offs between cost and performance is essential for guiding both research and commercialization efforts in this field [69].

(a) **Key Cost Factors**

The synthesis of carbon membranes involves several cost-intensive components. First, the raw materials used—such as graphite, polymers, or biomass—contribute significantly to the overall expense, especially when high-purity precursors are required for producing membranes with consistent and desirable properties [70]. Second, energy consumption is a major cost factor, as carbonization and activation processes typically involve sustained high temperatures and controlled atmospheres, leading to substantial energy usage [71]. Finally, equipment costs can be prohibitive. Large-scale synthesis methods such as Chemical Vapor Deposition (CVD) or pyrolysis require advanced reactors, high-temperature furnaces, and sometimes vacuum or inert gas systems, all of which demand significant capital investment [72].

(b) **Cost Reduction Strategies**

To address these cost barriers, several strategies have been proposed and implemented. One effective approach is the use of cheap precursors, such as biomass waste or recycled polymers [73]. These materials not only reduce raw material costs but also align with sustainable production goals [74].

Another strategy involves process optimization, where reaction temperatures, heating rates, and processing times are adjusted to minimize energy consumption without compromising membrane quality [75]. Additionally, one-pot synthesis or template-free methods are gaining attention for their potential to streamline production by eliminating complex intermediate steps, thereby reducing both time and cost [76].

(c) **Cost vs. Performance Trade-Offs**

Despite the push for cost reduction, it is important to consider the performance trade-offs that often accompany more economical choices [77]. For example, ultrathin porous membranes such as those based on graphene oxide offer exceptional selectivity and separation efficiency, making them suitable for high-precision applications like water purification or gas separation [78]. However, these membranes are significantly more expensive to produce. In contrast, activated carbon-based membranes are much more affordable and can be produced from readily available materials, but may fall short in terms of stability and selectivity [79]. Improving the performance of such cost-effective membranes through surface modification or composite formation remains an active area of research [80].

# 3 Fabrication Techniques and Pore Structure Engineering for Enhanced Selectivity

Fabrication techniques are crucial for controlling the pore structure of membranes to enhance their selectivity. By carefully manipulating the manufacturing process, researchers can fine-tune a membrane's ability to separate specific molecules or particles. One key method for achieving this is phase inversion [81].

## *3.1 Phase Inversion*

Phase inversion is a widely used membrane fabrication technique where a polymer undergoes a controlled transformation from a liquid to a solid state [82]. The process relies on liquid-liquid demixing, where a single-phase polymer solution separates into two distinct liquid phases [83]. One of these phases is rich in polymer and solidifies to form the final solid membrane matrix, while the other is rich in the non-solvent and creates the membrane's porous structure [84]. By precisely controlling the initial conditions of this phase transition, it's possible to produce both porous and non-porous membranes with a desired morphology [85]. This method is highly versatile and allows for the fine-tuning of pore size and distribution, which directly impacts the membrane's selectivity and permeability [86].

## 4 Surface Modification for Improved Compatibility in Various Applications

The surface of a carbon membrane can be modified to improve its performance in specific applications, particularly by enhancing its compatibility with different feed solutions and reducing fouling [87]. This involves altering the surface chemistry without compromising the membrane's bulk properties [88]. Techniques include plasma treatment, chemical grafting, and coating with a thin, functionalized layer [89]. For example, a hydrophobic carbon membrane can be made hydrophilic by grafting polar functional groups onto its surface [90]. This makes it more suitable for filtering aqueous solutions, as it can form a hydration layer that repels hydrophobic foulants, thus improving flux and long-term stability [91].

## 5 Challenges in Large-Scale Production and Cost Efficiency

Despite their exceptional performance, carbon membranes face significant hurdles in moving from laboratory research to industrial-scale production [92]. One major challenge is material consistency. The properties of the final product are highly sensitive to the type and quality of the carbon precursors used, and variations in raw materials can lead to inconsistencies in structure, porosity, and conductivity [93]. This problem becomes more pronounced as production volume increases, making quality control difficult [94].

Another barrier is the complexity of the synthesis processes. Methods like chemical vapor deposition (CVD) and templating techniques are effective in the lab but are difficult to replicate on a larger scale [64]. These techniques require specialized equipment and precise control of reaction parameters, which is costly and difficult to manage in a continuous production environment [95]. Finally, equipment and infrastructure limitations are a significant barrier. Many facilities lack the advanced reactors, high-temperature furnaces, or vacuum systems needed for industrial-scale production [67]. Retrofitting or building new production lines requires substantial investment.

The high cost of carbon membranes is also a major issue, driven by expensive raw materials, high energy consumption for high-temperature processes, and the cost of specialized equipment [96]. To address this, researchers are exploring cost reduction strategies like using cheap precursors, such as biomass waste or recycled polymers [73]. However, these economical choices can lead to performance trade-offs, such as lower stability and selectivity.

# 6 Conclusion and Prospective Applications

In conclusion, the synthesis of high-performance carbon-based membranes relies on a variety of innovative techniques that allow for precise control over structure and properties [97]. Methods such as pyrolysis, CVD, and sol-gel processing enable the creation of materials with tailored pore sizes, surface areas, and chemical compositions for specific applications in gas separation, water purification, and other filtration processes [98]. While these materials show immense promise, their widespread adoption is currently limited by significant challenges in large-scale production, including issues with material consistency, the complexity of synthesis processes, and high production costs [99].

Future research must focus on overcoming these barriers. This includes developing new, cost-effective synthesis methods and utilizing sustainable precursors to reduce reliance on expensive materials [100]. Further innovation in surface modification and pore structure engineering is also needed to enhance the performance and durability of carbon membranes, making them more competitive with established polymeric technologies [101]. By addressing these key challenges, carbon membranes can move from being a specialized, high-value material to a viable and scalable solution for a wide range of industrial applications.

## References


1. Mauter, M.S., Elimelech, M.: Environmental applications of carbon-based nanomaterials. Environ. Sci. Technol. **42**(16), 5843–5859 (2008)
2. Singh, V.N.: Chemical Methods for Processing Nanomaterials. CRC Press, Boca Raton, FL (2021)
3. Kong, D., et al.: Rational design of carbon-rich materials for energy storage and conversion. Adv. Mater. **31**(45), 1804973 (2019)
4. Morosanu, C.E.: Thin Films by Chemical Vapour Deposition, vol. 7. Elsevier (2016)
5. Mazzocchetti, L., et al.: Validation of carbon fibers recycling by pyro-gasification: the influence of oxidation conditions to obtain clean fibers and promote fiber/matrix adhesion in epoxy composites. Compos. A: Appl. Sci. Manuf. **112**, 504–514 (2018)
6. Hackbarth, H.G.: Tailorable ceramic nanocomposites through modular polymer chemistry (2024)
7. Sazali, N.: A review of the application of carbon-based membranes to hydrogen separation. J. Mater. Sci. **55**(25), 11052–11070 (2020)
8. Xia, C., et al.: Evolution and synthesis of carbon dots: from carbon dots to carbonized polymer dots. Adv. Sci. **6**(23), 1901316 (2019)
9. Manawi, Y.M., et al.: A review of carbon nanomaterials' synthesis via the chemical vapor deposition (CVD) method. Materials. **11**(5), 822 (2018)
10. Siddiqui, M.T.H., et al.: Synthesis of magnetic carbon nanocomposites by hydrothermal carbonization and pyrolysis. Environ. Chem. Lett. **16**(3), 821–844 (2018)
11. Danks, A.E., Hall, S.R., Schnepp, Z.: The evolution of 'sol–gel'chemistry as a technique for materials synthesis. Mater. Horiz. **3**(2), 91–112 (2016)
12. Tay, Y.S.: New organic precursors for carbon material: precursor development and carbonization process (2021)

13. Ward, J.W.: Thin Film Catalyst and Substrate Interactions on the CCVD (Catalytic Chemical Vapor Deposition) Growth of Carbon Nanotubes. Rensselaer Polytechnic Institute (2002)
14. Zhang, P., Zhu, H., Dai, S.: Porous carbon supports: recent advances with various morphologies and compositions. ChemCatChem. **7**(18), 2788–2805 (2015)
15. Pal, N., Bhaumik, A.: Soft templating strategies for the synthesis of mesoporous materials: inorganic, organic–inorganic hybrid and purely organic solids. Adv. Colloid Interf. Sci. **189**, 21–41 (2013)
16. Salleh, W.N.W., et al.: Precursor selection and process conditions in the preparation of carbon membrane for gas separation: a review. Sep. Purif. Rev. **40**(4), 261–311 (2011)
17. Marchetti, P., et al.: Molecular separation with organic solvent nanofiltration: a critical review. Chem. Rev. **114**(21), 10735–10806 (2014)
18. Bernardo, P., Drioli, E., Golemme, G.: Membrane gas separation: a review/state of the art. Ind. Eng. Chem. Res. **48**(10), 4638–4663 (2009)
19. Baker, R.W.: Research needs in the membrane separation industry: looking back, looking forward. J. Membr. Sci. **362**(1–2), 134–136 (2010)
20. Chong, H., et al.: A review on membrane fabrication: structure, properties and performance relationship. J. Appl. Membrane Sci. Technol. **29**(1), 73–97 (2025)
21. Asadizanjani, N., Reddy Kottur, H., Dalir, H.: CVD in semiconductor packaging. In: Introduction to Microelectronics Advanced Packaging Assurance, pp. 41–60. Springer (2025)
22. Wang, M., et al.: CVD polymers for devices and device fabrication. Adv. Mater. **29**(11), 1604606 (2017)
23. Jones, A.C., Hitchman, M.L.: Overview of chemical vapour deposition. In: Chemical Vapour Deposition: Precursors, Processes and Applications, vol. 1, pp. 1–36 (2009)
24. Barlow III, F.D., Elshabini, A.: Ceramic Interconnect Technology Handbook. CRC Press (2018)
25. Panjan, P., et al.: Review of growth defects in thin films prepared by PVD techniques. Coatings. **10**(5), 447 (2020)
26. Zong, T., et al.: A new smog chamber system for atmospheric multiphase chemistry study: design and characterization. Atmos. Meas. Tech. **16**(15), 3679–3692 (2023)
27. Denlinger, D., et al.: Thin film microcalorimeter for heat capacity measurements from 1.5 to 800 K. Rev. Sci. Instrum. **65**(4), 946–959 (1994)
28. Mountziaris, T.J., Jensen, K.F.: Gas-phase and surface reaction mechanisms in MOCVD of GaAs with trimethyl-gallium and arsine. J. Electrochem. Soc. **138**(8), 2426 (1991)
29. Vir Singh, M., Kumar Tiwari, A., Gupta, R.: Catalytic chemical vapor deposition methodology for carbon nanotubes synthesis. ChemistrySelect. **8**(32), e202204715 (2023)
30. Ghaemi, F., et al.: Synthesis of carbon nanomaterials using catalytic chemical vapor deposition technique. In: Synthesis, Technology and Applications of Carbon Nanomaterials, pp. 1–27. Elsevier (2019)
31. Do, P.T.M., Nguyen, L.X.: A review of thermochemical decomposition techniques for biochar production. Environ. Dev. Sustain., 1–57 (2024)
32. Zhang, B., Jiang, Y., Balasubramanian, R.: Synthesis, formation mechanisms and applications of biomass-derived carbonaceous materials: a critical review. J. Mater. Chem. A. **9**(44), 24759–24802 (2021)
33. Rogachev, A.S., Mukasyan, A.S.: Combustion for Material Synthesis. CRC Press (2014)
34. Ronsse, F., Nachenius, R.W., Prins, W.: Carbonization of biomass. In: Recent Advances in Thermo-chemical Conversion of Biomass, pp. 293–324. Elsevier (2015)
35. Ibrahim, H.A.-H.: Introductory chapter: pyrolysis. In: Recent Advances in Pyrolysis, p. 1 (2020)
36. Borucka, M., et al.: Identification of volatile and semi-volatile organic compounds emitted during thermal degradation and combustion of triadimenol. J. Therm. Anal. Calorim. **139**(2), 1493–1506 (2020)
37. Sierra-Jimenez, V., et al.: Atomistic modeling of lignocellulosic and carbonaceous fuels and their pyrolysis reactions: a review. Energy Fuel. **37**(23), 18408–18440 (2023)

38. Loutzenhiser, P.G., Schrader, A.J.: Exploiting reversible reduction/oxidation reactions for solar thermochemical energy storage in two-step cycles. Annu. Rev. Heat Transf., 25 (2022)
39. Agarwal, N., et al.: An overview of carbon-carbon composite materials and their applications. Front. Mater. **11**, 1374034 (2024)
40. Devi, M., Rawat, S., Sharma, S.: A comprehensive review of the pyrolysis process: from carbon nanomaterial synthesis to waste treatment. Oxford Open Mater. Sci. **1**(1), itab014 (2021)
41. Hamm, J.B., et al.: Experimental and computational analysis of carbon molecular sieve membrane formation upon polyetherimide pyrolysis. Carbon. **119**, 21–29 (2017)
42. Steel, K.M.: Carbon Membranes for Challenging Gas Separations. The University of Texas at Austin (2000)
43. Chan, K.F., et al.: Recent developments in carbon nanotubes-reinforced ceramic matrix composites: a review on dispersion and densification techniques. Crystals. **11**(5), 457 (2021)
44. Corriu, R., Anh, N.T.: Molecular Chemistry of Sol-Gel Derived Nanomaterials. John Wiley & Sons (2009)
45. Sharma, M., Pathak, M., Kapoor, P.N.: The sol-gel method: pathway to ultrapure and homogeneous mixed metal oxide nanoparticles. Asian J. Chem. **30**(7), 1405–1412 (2018)
46. Landfester, K.: Synthesis of colloidal particles in miniemulsions. Annu. Rev. Mater. Res. **36**(1), 231–279 (2006)
47. Hench, L.L., West, J.K.: The sol-gel process. Chem. Rev. **90**(1), 33–72 (1990)
48. Nwankwo, J.A., et al.: Microemulsion gel systems: formulation, stability studies, biopolymer interactions, and functionality in food product development. Compr. Rev. Food Sci. Food Saf. **24**(2), e70110 (2025)
49. Şahin, İ., et al.: Kinetics of supercritical drying of gels. Gels. **4**(1), 3 (2017)
50. Job, N., et al.: Carbon aerogels, cryogels and xerogels: influence of the drying method on the textural properties of porous carbon materials. Carbon. **43**(12), 2481–2494 (2005)
51. Zelinski, B., Uhlmann, D.: Gel technology in ceramics. J. Phys. Chem. Solids. **45**(10), 1069–1090 (1984)
52. Ramírez, C., et al.: Applications of ceramic/graphene composites and hybrids. Materials. **14**(8), 2071 (2021)
53. Ivanova, O., Williams, C., Campbell, T.: Additive manufacturing (AM) and nanotechnology: promises and challenges. Rapid Prototyp. J. **19**(5), 353–364 (2013)
54. Sheoran, K., et al.: Recent advances of carbon-based nanomaterials (CBNMs) for wastewater treatment: synthesis and application. Chemosphere. **299**, 134364 (2022)
55. Forgie, J.R., et al.: Electrospinning of high-performance nanofibres: state of the art and insights into the path forward. Appl. Sci. **13**(22), 12476 (2023)
56. Islam, M.A., et al.: Melt electrospinning for membrane fabrication. In: Electrospun and Nanofibrous Membranes, pp. 27–51. Elsevier (2023)
57. Hösel, M., Dam, H.F., Krebs, F.C.: Development of lab-to-fab production equipment across several length scales for printed energy technologies, including solar cells. Energ. Technol. **3**(4), 293–304 (2015)
58. Baig, N., Kammakakam, I., Falath, W.: Nanomaterials: a review of synthesis methods, properties, recent progress, and challenges. Mater. Adv. **2**(6), 1821–1871 (2021)
59. Baxendale, I.R., et al.: Achieving continuous manufacturing: technologies and approaches for synthesis, workup, and isolation of drug substance. May 20–21, 2014 Continuous Manufacturing Symposium. J. Pharm. Sci. **104**(3), 781–791 (2015)
60. Bazaka, K., Jacob, M.V., Ostrikov, K.: Sustainable life cycles of natural-precursor-derived nanocarbons. Chem. Rev. **116**(1), 163–214 (2016)
61. Gualtieri, M.L., et al.: Thermal conductivity of fired clays: effects of mineralogical and physical properties of the raw materials. Appl. Clay Sci. **49**(3), 269–275 (2010)
62. Mitra, A.: Fundamentals of Quality Control and Improvement. John Wiley & Sons (2016)
63. Levashov, E., et al.: Self-propagating high-temperature synthesis of advanced materials and coatings. Int. Mater. Rev. **62**(4), 203–239 (2017)
64. Cai, Z., et al.: Chemical vapor deposition growth and applications of two-dimensional materials and their heterostructures. Chem. Rev. **118**(13), 6091–6133 (2018)

65. Charpentier, J., McKenna, T.: Managing complex systems: some trends for the future of chemical and process engineering. Chem. Eng. Sci. **59**(8–9), 1617–1640 (2004)
66. Asch, M., et al.: Big data and extreme-scale computing: pathways to convergence-toward a shaping strategy for a future software and data ecosystem for scientific inquiry. Int. J. High Perform. Comput. Appl. **32**(4), 435–479 (2018)
67. Bramsiepe, C., et al.: Low-cost small scale processing technologies for production applications in various environments—Mass produced factories. Chem. Eng. Process. Process Intensif. **51**, 32–52 (2012)
68. Pacheco-Torgal, F.: Introduction to cost-effective energy-efficient methods for refurbishment and retrofitting of buildings. In: Cost-Effective Energy-Efficient Methods for Refurbishment and Retrofitting of Buildings, pp. 1–19. Elsevier (2025)
69. Kaplan, R.S., Cooper, R.: Cost & Effect: Using Integrated Cost Systems to Drive Profitability and Performance. Harvard Business Press (1998)
70. Yan, Y., et al.: Synthesis of graphene: potential carbon precursors and approaches. Nanotechnol. Rev. **9**(1), 1284–1314 (2020)
71. Song, C.: Global challenges and strategies for control, conversion and utilization of CO2 for sustainable development involving energy, catalysis, adsorption and chemical processing. Catal. Today. **115**(1–4), 2–32 (2006)
72. Bonney III, M.M.: Hot Wire Chemical Vapor Deposition of Tungsten Oxides for Photocatalytic Hydrogen Production from Water. University of Arkansas at Little Rock (2023)
73. Mohanty, A.K., et al.: Sustainable polymers. Nat. Rev. Methods Primers. **2**(1), 46 (2022)
74. Hegab, H., et al.: Toward sustainable future: strategies, indicators, and challenges for implementing sustainable production systems. Sustain. Mater. Technol. **36**, e00617 (2023)
75. Drioli, E., et al.: Process intensification strategies and membrane engineering. Green Chem. **14**(6), 1561–1572 (2012)
76. Zhu, M., et al.: A review of synthetic approaches to hollow nanostructures. Mater. Chem. Front. **5**(6), 2552–2587 (2021)
77. Gibson, R.B.: Avoiding sustainability trade-offs in environmental assessment. Impact Assess. Proj. Apprais. **31**(1), 2–12 (2013)
78. Zhang, S., et al.: Ultrathin membranes for separations: a new era driven by advanced nanotechnology. Adv. Mater. **34**(21), 2108457 (2022)
79. Fan, X., Wei, G., Quan, X.: Carbon nanomaterial-based membranes for water and wastewater treatment under electrochemical assistance. Environ. Sci. Nano. **10**(1), 11–40 (2023)
80. Saleh, E.A.M., et al.: Recent advances in thin film composite (TFC) membranes development: materials and modification methods. Environ. Sci. Water Res. Technol. (2025)
81. Cheng, X., Bae, J.: Recent advancements in fabrication, separation, and purification of hierarchically porous polymer membranes and their applications in next-generation electrochemical energy storage devices. Polymers. **16**(23), 3269 (2024)
82. Ren, J., Wang, R.: Preparation of polymeric membranes. In: Membrane and Desalination Technologies, pp. 47–100. Springer (2010)
83. van Vliet, R.E., et al.: Dynamics of liquid–liquid demixing: mesoscopic simulations of polymer solutions. Fluid Phase Equilib. **201**(1), 67–78 (2002)
84. Dong, X., et al.: Polymers and solvents used in membrane fabrication: a review focusing on sustainable membrane development. Membranes. **11**(5), 309 (2021)
85. Stucki, M., Loepfe, M., Stark, W.J.: Porous polymer membranes by hard templating–a review. Adv. Eng. Mater. **20**(1), 1700611 (2018)
86. Wang, J., et al.: Nanoscale tailor-made membranes for precise and rapid molecular sieve separation. Nanoscale. **9**(9), 2942–2957 (2017)
87. Rana, D., Matsuura, T.: Surface modifications for antifouling membranes. Chem. Rev. **110**(4), 2448–2471 (2010)
88. Miller, D.J., et al.: Surface modification of water purification membranes. Angew. Chem. Int. Ed. **56**(17), 4662–4711 (2017)
89. Oehr, C., et al.: Plasma grafting—a method to obtain monofunctional surfaces. Surf. Coat. Technol. **116**, 25–35 (1999)

90. Ayyaru, S., Ahn, Y.-H.: Application of sulfonic acid group functionalized graphene oxide to improve hydrophilicity, permeability, and antifouling of PVDF nanocomposite ultrafiltration membranes. J. Membr. Sci. **525**, 210–219 (2017)
91. Zhang, R., et al.: Antifouling membranes for sustainable water purification: strategies and mechanisms. Chem. Soc. Rev. **45**(21), 5888–5924 (2016)
92. Mollahosseini, A., et al.: Polymeric membranes in carbon capture, utilization, and storage: current trends and future directions in decarbonization of industrial flue gas and climate change mitigation. Energy Environ. Sci. **18**(11), 5025–5092 (2025)
93. Wang, Y., et al.: Carbonaceous materials in structural dimensions for advanced oxidation processes. Chem. Soc. Rev. **54**, 2436–2482 (2025)
94. Hackman, J.R., Wageman, R.: Total quality management: empirical, conceptual, and practical issues. Adm. Sci. Q., 309–342 (1995)
95. Gutmann, B., Cantillo, D., Kappe, C.O.: Continuous-flow technology—a tool for the safe manufacturing of active pharmaceutical ingredients. Angew. Chem. Int. Ed. **54**(23), 6688–6728 (2015)
96. Ismail, A.F., David, L.: A review on the latest development of carbon membranes for gas separation. J. Membr. Sci. **193**(1), 1–18 (2001)
97. Dutta, V., et al.: Bio-inspired synthesis of carbon-based nanomaterials and their potential environmental applications: a state-of-the-art review. Inorganics. **10**(10), 169 (2022)
98. Mostafavi, A.H., et al.: Advances in surface modification and functionalization for tailoring the characteristics of thin films and membranes via chemical vapor deposition techniques. J. Appl. Polym. Sci. **140**(15), e53720 (2023)
99. Lehmann, T., et al.: Large-scale metal additive manufacturing: a holistic review of the state of the art and challenges. Int. Mater. Rev. **67**(4), 410–459 (2022)
100. Mariotti, N., et al.: Recent advances in eco-friendly and cost-effective materials towards sustainable dye-sensitized solar cells. Green Chem. **22**(21), 7168–7218 (2020)
101. Li, L., Duan, Y.: Engineering polymer-based porous membrane for sustainable lithium-ion battery separators. Polymers. **15**(18), 3690 (2023)

# Understanding Mass Transport in Carbon-Based Membranes

**Ferdy Subandi, Abisha Jaumar Widjanarko, Shabira, Januar Widakdo, and Yulia Mariana Tesa Ayudia Putri**

## 1 Introduction

Mass transport in carbon-based membranes is a fundamental aspect that determines the performance and efficiency of applications such as gas separation [1–7], water purification [6, 8–10], and nanofiltration [11–13]. The unique pore structure characteristics of carbon membranes, including pore size, shape, and distribution, directly influence the mechanisms of fluid molecule transport through the membrane [14, 15]. By understanding various transport modes occurring from the micro to macro scale, researchers and engineers can design membranes that are more selective and efficient [16].

The pore structures of carbon membranes can be classified into three main categories: micropores (<2 nm), mesopores (2–50 nm), and macropores (>50 nm) [17, 18]. Each category is associated with distinct mass transport mechanisms, including constrained molecular diffusion, Knudsen diffusion, and viscous flow [19]. The combination of pore size and fluid properties (gas or liquid) determines the dominant transport pathway within the system [20]. This chapter systematically discusses these transport mechanisms and their relevance to the design and development of functional carbon membranes.

The objective of this chapter is to provide a comprehensive understanding of mass transport mechanisms in carbon-based membranes, focusing on models of gas

F. Subandi · A. J. Widjanarko · Shabira · J. Widakdo
Department of Physics, Faculty of Mathematics and Natural Sciences, Universitas Indonesia, Depok, Indonesia

Yulia Mariana Tesa Ayudia Putri (✉)
Department of Chemistry, Faculty of Mathematics and Natural Sciences, Universitas Indonesia, Depok, Indonesia
e-mail: Yulia.mariana@sci.ui.ac.id



H. Abdullah, J. Widakdo (eds.), *Membrane Physics of Carbon-Based Materials*, Engineering Materials, https://doi.org/10.1007/978-981-95-5605-2_3

and liquid transport, molecular diffusion studies through membrane pores using theoretical and experimental approaches, the effects of structural and compositional modifications on membrane permeability and selectivity, and numerical simulations and theoretical calculations for understanding fluid transport in carbon membranes. Each subchapter will explore these aspects in detail, supported by data and references from recent scientific literature.

## 2 Models of Gas and Liquid Transport in Carbon-Based Membranes

Models of gas and liquid transport in carbon-based membranes largely depend on the classification of pore structures, which define the diffusion pathways for fluids. Each pore type, micropores, mesopores, and macropores, provides a distinct transport environment influenced by molecular size, temperature, pressure, and membrane surface characteristics [21].

In microporous structures, the movement of gas or liquid molecules is highly constrained [22]. Gases primarily undergo constrained molecular diffusion, while liquids experienced both restricted diffusion and surface transfer due to adsorption interactions between the molecules and the pore walls [23]. In mesopores structures, gas movement occurs mainly through Knudsen diffusion, where molecular motion is more frequently interrupted by pore wall collisions than by intermolecular collisions [24]. For liquids, viscous (laminar) flow dominates, following classical fluid mechanics principles [25]. In macroporous structures, both gases and liquids are primarily transported through viscous flow due to the large pore size that allows relatively free fluid movement with minimal resistance [26].

These transport models form the foundation for designing carbon membranes for specific applications such as hydrogen separation, organic contaminant removal, and desalination processes [27]. The selection of pore structure and control of diffusion mechanisms are crucial to optimizing membrane performance in terms of selectivity and permeability [28] (Table 1).

### 2.1 *Micropore Transport*

Transport in micropores (pore diameter $< 2$ nm) exhibits unique characteristics because the pore size is comparable to the molecular size [32]. Under these conditions, the difference between gas and liquid molecules becomes negligible, as both experience strong attractive forces from the pore walls [33]. Consequently, the molecules cannot move freely as in open space but must transfer from one vacant position to another, analogous to hopping from site to site [34]. This phenomenon is referred to as "hopping diffusion," and it is commonly described using the Langmuir

**Table 1** Classification of gas and liquid permeation through the membrane pore

| Membrane pores | Gas transport (dominant mechanisms) | Liquid transport (dominant mechanisms) |
|---|---|---|
| **Micropore** (≈<2 nm) | Constrained **molecular (configurational) diffusion; surface (adsorbate) diffusion** often important when adsorption is strong | Constrained molecular diffusion (sorption–diffusion) and **surface transfer**/transport along adsorbed film; capillary condensation can appear in some cases [29] |
| **Mesopore** (≈2–50 nm) | **Knudsen diffusion** (molecule–wall collisions dominate) to **transition regime** (Knudsen + molecular) depending on mean free path and pore diameter; slip/transition effects may appear | **Viscous (Poiseuille) flow** begins to contribute; transport may be mixed (surface diffusion + viscous flow) depending on wetting and pore connectivity [30] |
| **Macropore** (≳50 nm) | **Viscous (continuum/Poiseuille) flow** dominates (intermolecular collisions ≫ molecule–wall collisions); Darcy-like flow often applicable | **Viscous (Poiseuille/Darcy) flow** dominates; bulk liquid flow, possibly with inertial effects in very large pores [31] |

lattice model, which represents a lattice system in which each site can either be vacant or occupied by a single molecule [35].

Molecules experience attraction from the pore walls and move according to the hopping diffusion mechanism, similar to the vacancy mechanism in solid-state systems, where movement occurs from one vacant site to another. Micropores can thus be modeled as a network of sites (Langmuir lattice) that may be either vacant or occupied by molecules. Molecular movement depends on the availability of vacant sites and on the strength of molecular interaction with the pore walls.

Species flux is formulated as follow:

$$j_\iota = -f_{L1} \cdot b_i^0 \cdot c^{\text{tot}} \cdot RT \left[ (1-\theta_2)\frac{\partial \theta_1}{\partial z} + \theta_1 \frac{\partial \theta_2}{\partial z} \right]$$

where $j_\iota$ is the species flux, $f_{L1}$ the non-equilibrium correction factor, $b_i^0$ the mechanical mobility, $c^{\text{tot}}$ the total concentration of sites, $R$ the gas constant, $T$ the absolute temperature, and $\theta$ the site occupancy fraction. The species flux described can be simplified according to its transport mechanism, as follow:

For species fluxes with low site occupancy $\theta \ll 1$, the molecules do not interfere with each other and follow Fick's Law. The equation simplifies to:

$$j_\iota = -b_i^0 \cdot c^{tot} \cdot RT \frac{\partial \theta_i}{\partial z} = -D_i \cdot \frac{\partial \theta_i}{\partial z}$$

$$D_i = b_i^0 \cdot RT$$

If Henry's Law applies:

$$\theta_i = K_i^H \cdot \frac{p_i}{p^0}$$

Then, the diffusion permeability is expressed as:

$$f_\iota^1 = \frac{b_\iota^0\, C^{tot}\, RT \dfrac{\partial \theta_i}{\partial z}}{P_{\iota f} - P_{\iota P}} = \frac{C^{tot}\, D_i\, K_i^H}{P^0\, X}$$

where $P_{\iota f}$ and $P_{\iota P}$ are the feed and permeate partial pressures, $X$ is the membrane thickness, and $P^0$ is the standard pressure.

In contrast, when site occupancy is high (the molecules occupy more than 50% of available sites), a percolation network is formed. This is characterized by strong molecular affinity toward the pore walls, low mobility, and inhibition of other molecules' movement within the pore system.

## 2.2 Knudsen Transport

The Knudsen transport model is a form of gas diffusion that becomes dominant in porous systems when the pore diameter is smaller than the mean free path ($\lambda$) of the gas molecules. This condition is typically observed in mesoporous membranes (pore sizes ranging from 2–50 nm) or under low-pressure conditions. In such scenarios, gas molecules collide more frequently with the pore walls than with each other, making wall collisions the primary mode of transport [34].

Knudsen transport occurs when:

$$\lambda > \Phi_p$$

where $\lambda$ is the mean free path of the gas molecules and $\Phi_p$ the pore diameter of the membrane.

Under this regime, intermolecular collisions are infrequent, and gas movement is largely governed by the geometry of the pores and the physical characteristics of the gas molecules. The gas permeance for species through a membrane under Knudsen conditions is given by:

The gas permeance for species through a membrane under Knudsen conditions is given by:

$$f_\iota^{kn} = \frac{\varphi_p \Phi_p}{3\tau\, X} \sqrt{\frac{8}{\pi\, RT\, M_\iota}}$$

where the following value $\varphi_p$ is the volume fraction of porosity, is the tortuosity of the pores, and $M_\iota$ is the molecular weight of the gas.

This equation shows that gas permeance increases with higher porosity and larger pore diameter, while it decreases with greater membrane thickness, higher tortuosity, and heavier molecular weight [11].

In gas separation applications, membrane performance is often assessed based on its selectivity for two gases. Knudsen selectivity between two gases and is expressed as:

$$\alpha_{i_2 i_1}^{Kn} = \sqrt{\frac{M_{i_2}}{M_{i_1}}}$$

This selectivity depends solely on the molecular weight ratio. For instance, for hydrogen ($H_2$, 2 g/mol) and nitrogen ($N_2$, 28 g/mol), the Knudsen selectivity is:

$$\alpha_{\frac{N_2}{H_2}}^{Kn} = \sqrt{\frac{2}{28}} = 0.267$$

Thus, gases with lower molecular weights diffuse faster through mesoporous membranes [28].

Although the Knudsen model is mathematically well-defined, it presents several limitations, particularly for selective separation purposes. Since selectivity is based purely on molecular weight, the mechanism is ineffective for separating gases with similar molecular weights or in cases where chemical interactions play a role. Moreover, Knudsen transport does not account for surface interactions, which may be significant in microporous membranes or under more complex operational conditions.

When the pore size approaches the size of the gas molecules ($\theta \rightarrow 1$), two notable phenomena may occur: The first is single-file diffusion, where the molecules move in a line without the ability to bypass each other inside the pore. The second being molecular correlation effects, with interactions between molecules within narrow pores can significantly reduce diffusion rates. Under such conditions, classical Knudsen theory becomes insufficient, and more advanced models like molecular dynamics simulations are required.

## 2.3 *Viscous Flow Transport*

Viscous transport is one of the primary mechanisms of fluid movement through the pores of a membrane. This mechanism occurs due to a pressure gradient between the two sides of the membrane, rather than a concentration gradient as seen in diffusion. Fluids under high pressure are driven through the membrane pores toward areas of lower pressure [34].

To describe this phenomenon, the Hagen–Poiseuille equation is commonly applied. This equation is valid for Newtonian fluid flow in small cylindrical capillaries under steady-state conditions. It has been adapted to represent the viscous permeance (flow rate per unit pressure and area) in membranes, as follows:

$$f_{\iota}^{visc} = \frac{\varphi_p \Phi_p^2}{32\tau X}$$

where $f_l^{visc}$ is the viscous permeance ($m^2$/Pa s), indicating the efficiency of viscous flow through the membrane. This equation indicates that viscous permeance increases with higher porosity and larger pore diameter, but decreases with increasing tortuosity and membrane thickness. Therefore, a membrane optimized for viscous transport would ideally possess large, numerous, straight pores and a thin structure [34].

In general, viscous flow is considered a non-selective transport mechanism, meaning that any molecule smaller than the pore size can pass through without discrimination based on size, molecular weight, or chemical properties. This characteristic distinguishes it from selective transport mechanisms such as molecular sieving or diffusion. Nevertheless, under specific conditions, viscous flow may still exhibit partial selectivity, due to the following effects:

The first is size exclusion effect, which is when the size of the solute molecules is close to the pore diameter, the flow field near the pore entrance may create an additional resistance that reduces the likelihood of those molecules entering the pore. This can lead to partial exclusion based on size. In practical terms, even though viscous flow is theoretically non-selective, large molecules may still be hindered due to spatial mismatches with the pore structure.

The second is space charge effect, which is for mesoporous membranes with charged pore walls, such as those used in electrolyte solutions, ions in the solution can interact with the electric field within the pores. If the pore diameter is smaller than the Debye length of the solution, the electrical field from the charged pore walls can repel like-charged ions. This phenomenon can result in ion retention through electrostatic repulsion, depending on the interaction between the ion charge and the surface charge of the pores. As a result, viscous flow in charged membranes may exhibit charge-based selectivity even if the molecular sizes are appropriate.

## 3 Solution-Diffusion Model for the Transport of Binary Gas Mixtures

For the experimental part, the diffusivities of binary mixtures were measured using the laboratory fabricated CMSM. For the theory part, we use developed kinetic theory for predict binary diffusivities from pure component diffusivities. The diffusivity measurement was made by the diffusion cell. The laboratory prepared CMSM was loaded between two chambers of the diffusion cell. Pure helium (He) gas was fed into the lower chamber on the permeate side while He gas carrying different concentrations of $CH_4$, $C_2H_6$ or $CH_4/C_2H_6$ mixture was fed to the upper chamber on the feed side of the diffusion cell. The pressures on both chambers were kept equal. Hence, the gas transport was solely by the diffusion. From the flow rate and the concentration of the permeant at the outlet of the permeate side stream, the permeant flux can be calculated. The flux measurement was done after waiting for 8 h, at least, to ensure the establishment of the steady state.

Figures 1 and 2 show the SEM pictures of the laboratory prepared CMSM and the substrate graphite. Figure 1 depicts that the CMSM coated on top of the graphite support is crack-free. The thickness of the CMSM layer is 15 μm. Figure 2 shows that the surface of the CMSM and the graphite support. The CMSM layer is much smoother with a roughness within 0.02 μm. The pore size in the graphite support is 5–10 μm.

The adsorption isotherms for $CH_4$ and $C_2H_6$ is given in Fig. 3. The data were fitted to the Sippe type isotherm:

$$\frac{q}{q_s} = \frac{bp^n}{1+bp^n}$$

where $p$ is the gas pressure, $q$ is the amount adsorbed and $qs$ is the saturated amount adsorbed. $b$ is the Langmuir constant and $n$ is isotherm constant. Regarding the single component gas diffusion, the flux can be written as:

$$J = -D\frac{\partial q}{\partial x}$$

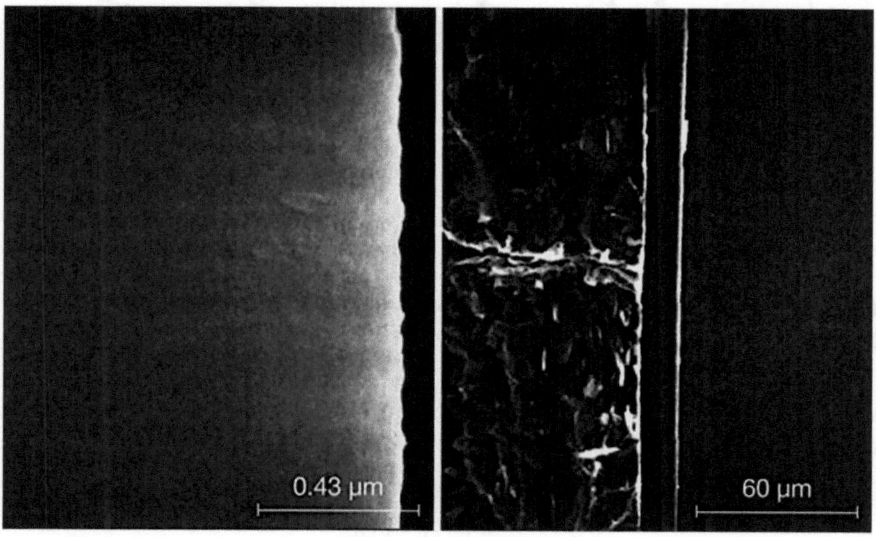


**Fig. 1** SEM cross-sectional images of (right) CMS layer formed on the surface of graphite substrate and (left) CMS layer and its surface [36]. Reproduced from ref. [36] with permission from American Chemical Society, copyright 1994

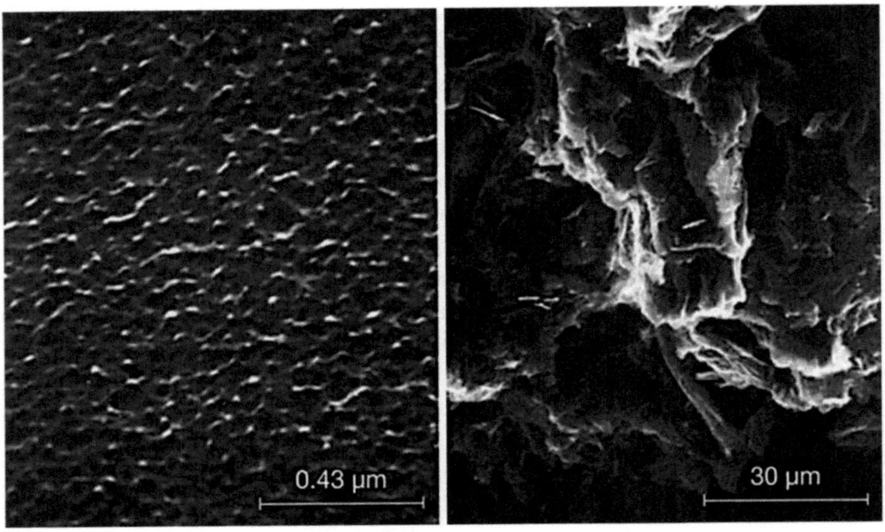


**Fig. 2** SEM images of (left) the surface of CMS layer, and (right) the surface of graphite substrate [36]. Reproduced from ref. [36] with permission from American Chemical Society, copyright 1994

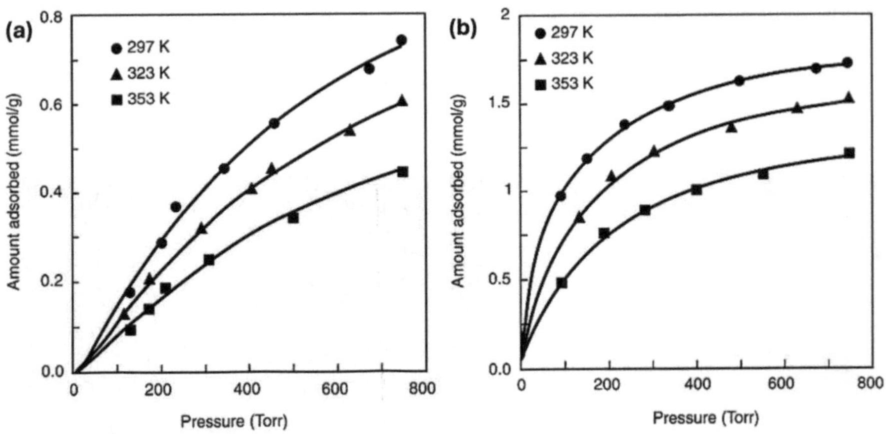


**Fig. 3** Equilibrium isotherms of $CH_4$ and $C_2H_6$ on CMS consecutively [36]. Reproduced from ref. [36] with permission from American Chemical Society, copyright 1994

where $D$ is the diffusivity and $x$ is distance. Diffusion in molecular sieves is strongly concentration dependent, and is given by:

$$D = \frac{D_0}{1-(1-\lambda)\theta}$$

where $\theta$ is the fractional saturation, $\theta = q/qs$, $D_0$ is the diffusivity at zero adsorption and $\lambda$ is interaction parameter. At the steady state the gas flux is constant. Then, we can substitute $D$ to the flux and integrate to achieve the result:

**Table 2** Single-component equilibrium and diffusivity parameters [34]

| Gas | Temperature (K) | qs (mmol/g) | $b \times 10^4$ (1/Torr) | $n$ | $D_0 \times 10^8$ (cm$^2$/s) | $\lambda$ | $D \times 10^8$ (cm$^2$/s) |
|---|---|---|---|---|---|---|---|
| $CH_4$ | 297 | 1.145 | 4.91 | 1.24 | 1.695 | 0.056 | 2.379 |
| | 323 | 1.051 | 4.51 | 1.21 | 2.429 | 0.196 | 3.047 |
| | 353 | 0.797 | 4.21 | 1.22 | 3.428 | 0.280 | 4.445 |
| $C_2H_6$ | 297 | 2.072 | 20.0 | 0.84 | 0.146 | 0 | 0.293 |
| | 323 | 1.821 | 53.0 | 1.03 | 0.188 | 0 | 0.343 |
| | 353 | 1.549 | 49.4 | 0.99 | 0.315 | 0 | 0.524 |

$$J = -\frac{D_0}{\Delta x}\frac{q_s}{(1-\lambda)}\ln\left[\frac{q_s-(1-\lambda)q_L}{q_s-(1-\lambda)q_H}\right]$$

where $\Delta x$ is the thickness of CMS layer. The subscripts $H$ and $L$ are for the higher and lower chamber. Since $qH \gg qL$, the equation is further reduced to:

$$J = -\frac{D_0}{\Delta x}\frac{q_s}{(1-\lambda)}\ln\left[1-(1-\lambda)\frac{bp^n}{1+bp^n}\right]$$

Table 2 shows the parameters involved in Sippe type isotherm and diffusion in molecular sieves at different temperatures. When an assumption is made that the diffusivity is not concentration dependent, the flux becomes (Using Taylor series):

$$J = \frac{D}{\Delta x}\frac{q_s bp^n}{(1+bp^n)}\ln$$

The experimental $CH_4$ and $C_2H_6$ fluxes are shown in Figs. 4 and 5. The symbols are the experimental data. The solid lines are calculated results from the flux with diffusivity is concentration dependent using the parameters given in Table 2. The broken lines are the best fit data based on flux equation without concentration dependent.

Flux equation with concentration dependent gives a better result in fitting than flux equation without concentration dependent, meaning the effect of concentration on the diffusivity can't be ignored.

The parameter $\lambda$ defined for a specific molecule A is given as:

$$\lambda_A = e^{-(E_{AV}-E_{AA})/RT}$$

$\epsilon_{AV}$ and $\epsilon_{AA}$ mean the bond energy of the molecule A and the vacant site and the bond energy of molecule A and the molecule A that already occupied the vacant site, respectively.

In Table 2, $\lambda$ for C2H6 is equal to zero at all temperatures, meaning that $\varepsilon AV$ is much greater than $\varepsilon AA$, while $\lambda$ for $CH_4$ decreases with decreasing temperature, meaning that the A–V bonding increases as the temperature decreases.

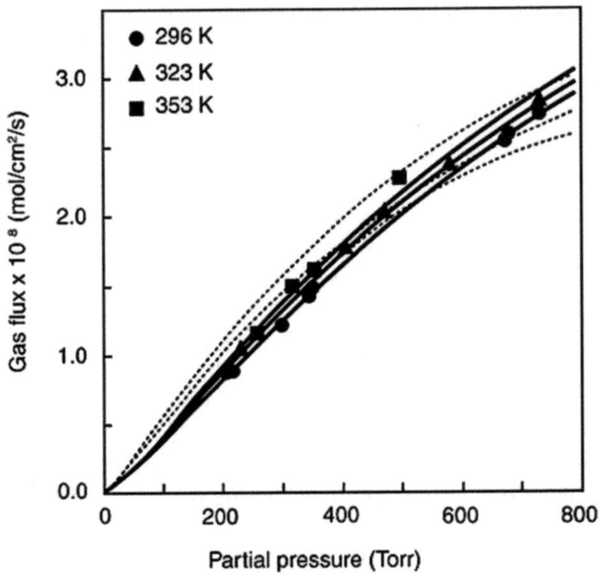


**Fig. 4** Steady-state flux of $CH_4$ through CMSM: solid curves are fitted with codependent diffusivity and dashed curves are with constant diffusivity [36]. Reproduced from ref. [36] with permission from American Chemical Society, copyright 1994

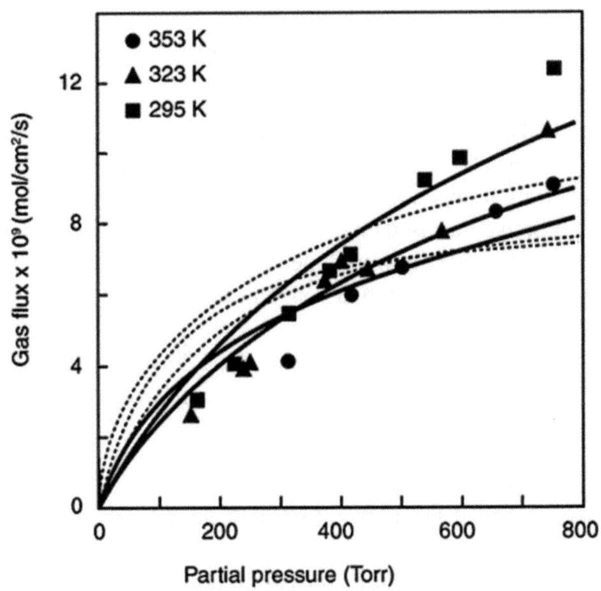


**Fig. 5** Steady-state flux of $C_2H_6$ through CMSM: solid curves are fitted with codependent diffusivity and dashed curves are with constant diffusivity [36]. Reproduced from ref. [36] with permission from American Chemical Society, copyright 1994

$D_0$ obtained at three temperatures can be given in the form of Arrhenius equation as:

$$D_0 = D_0^* e^{-E_a/RT}$$

where $D^*_0$ is the pre-exponential factor and $E$ is the activation energy of diffusion. The values of $D^*_0$ and $E_a$ are shown in Table 3.

For a binary system of components A and B, the fluxes are:

$$J_A = -D_{AA}\frac{\partial q_A}{\partial x} - D_{AB}\frac{\partial q_B}{\partial x}$$

$$J_B = -D_{BA}\frac{\partial q_A}{\partial x} - D_{BB}\frac{\partial q_B}{\partial x}$$

And the concentration dependent Fickian diffusivities are:

$$D_{AA} = D_{A0} \times \frac{1-(1-\lambda_{AB})\theta_B}{1-(1-\lambda_A)\theta_A-(1-\lambda_{AB})\theta_B}$$

$$D_{AB} = D_{A0} \times \frac{1-(1-\lambda_{AB})\theta_A}{1-(1-\lambda_A)\theta_A-(1-\lambda_{AB})\theta_B}$$

$$D_{BA} = D_{B0} \times \frac{1-(1-\lambda_{BA})\theta_B}{1-(1-\lambda_B)\theta_B-(1-\lambda_{BA})\theta_A}$$

$$D_{BB} = D_{B0} \times \frac{1-(1-\lambda_{BA})\theta_A}{1-(1-\lambda_B)\theta_B-(1-\lambda_{BA})\theta_A}$$

The cross-term $\lambda_{AB}$ is defined as:

$$\lambda_{AB} = e^{-(\epsilon_{AV}-\epsilon_{AB})/RT}$$

A similar expression is applicable for $\lambda_{BA}$.

The cross-term activation energies are given by

$$\epsilon_{AB} = \epsilon_{BA} = (\epsilon_{AA}\epsilon_{BB})^{1/2}$$

By substituting the diffusivity equations for binary system into the flux equations for binary system and integrating the flux equations over $q_A$ by keeping the other component at a constant average $\bar{q}_B$, the following equations are obtained.

**Table 3** Energetic parameters and pre-exponential factors

| Gas | $E_a$ (kcal/mol) | $D^*_0 \times 10^8$ (cm$^2$/s) |
|---|---|---|
| $CH_4$ | 2.51 | 1.22 |
| $C_2H_6$ | 2.97 | 0.21 |

$$J_i = -\frac{1}{\Delta x}\sum_{j=1}^{2}\left[\overline{D_{ij}}\left(\overline{q_{jout}} - q_{in}\right)\right], i = A\, and\, B$$

Finally, for the prediction based on a single-component system, the following equations are used.

$$J_A = -D_{AA}\frac{dq_A}{dx}$$

$$J_B = -D_{BB}\frac{dq_B}{dx}$$

$$D_{AA} = \frac{D_{A0}}{1-\left(1-\lambda_{AA}\right)\theta_A}$$

$$D_{BB} = \frac{D_{B0}}{1-\left(1-\lambda_{BB}\right)\theta_B}$$

Table 4 shows experimental flux data for the binary system together with predictions based on the binary system and single gas component system. It is obvious from Table 4 that prediction based on the binary system produces much better results.

# 4 Structure and Composition Modification Effect on Membrane Permeability and Selectivity

## 4.1 Structure Modification

The pore size and pore size distribution of carbon molecular sieve (CMS) membrane were modified by self-assisted deposition of carbon segment to enhance the separation performance of a gas pair. The poly(p-phenylene oxide) (PPO) casting

**Table 4** Comparison of prediction based on the binary system and the single component system [34]

| Feed gas mole fraction | | Experimental flux $\times 10^9$(mol/cm$^2$s) | | Flux predicted, binary theory $\times 10^9$(mol/cm$^2$s) | | Flux predicted, single component $\times 10^9$(mol/cm$^2$s) | |
|---|---|---|---|---|---|---|---|
| $CH_4$ | $C_2H_6$ | $CH_4$ | $C_2H_6$ | $CH_4$ | $C_2H_6$ | $CH_4$ | $C_2H_6$ |
| 0.611 | 0.021 | 18.8 | 0 | 17.5 | Small | 15.9 | Small |
| 0.138 | 0.58 | 3.18 | 11.6 | 1.67 | 6.07 | 0.989 | 5.94 |
| 0.093 | 0.43 | 1.99 | 9.6 | 1.19 | 5.36 | 0.737 | 5.27 |
| 0.465 | 0.317 | 13.6 | 5.06 | 8.53 | 3.97 | 5.66 | 3.58 |
| 0.349 | 0.625 | 8.04 | 10.4 | 4.78 | 5.85 | 2.82 | 5.47 |
| 0.218 | 0.125 | 5.49 | 2.26 | 4.87 | 2.56 | 3.67 | 2.42 |

dope was selected as the carbon segment-forming agent, coating the surface of the carbon membrane, followed by carbonization at 600 °C to form a modified carbon layer. The original carbon layer that used in here are polyamide (PI) and polyetherimide (PEI). The fabrications of the original also the modified one already explained in [37].

The TGA curves of the PI, PEI and PPO polymer membranes obtained from nitrogen atmosphere are shown in Fig. 6. For the PI polymer membrane, two thermal degradation stages can be observed (Fig. 6a). The beginning of the first stage encompasses the temperature range of 200–350 °C, with minimal weight loss of around 13–16 wt%, which can be attributed to the removal of residual NMP solvent present in the polymeric films [38]. The second stage of thermal degradation occurs from 500–600 °C, with a huge weight loss around 36 wt%, which is most likely due to the thermal decomposition, or breakage, of the main chains in the polymeric films [38]. At approximately 550 °C, maximum polymer decomposition occurs and significant physiochemical changes transpire on the carbon film due to the transformation of imide groups present in the polymeric membranes [38]. When the thermal degradation temperature reaches 700 °C, the weight loss slows down and reaches approximately 44 wt% of the total weight remaining on the membranes [38].

Figure 6b shows two similar stages of thermal degradation curves compared with Fig. 5. The minimal weight losses of the PEI polymer membrane at temperature ranges of 200–300 °C and 500–600 °C are approximately 5–16 and 43 wt%, respectively. At approximately 530 °C, the decomposition rate increases, which may be

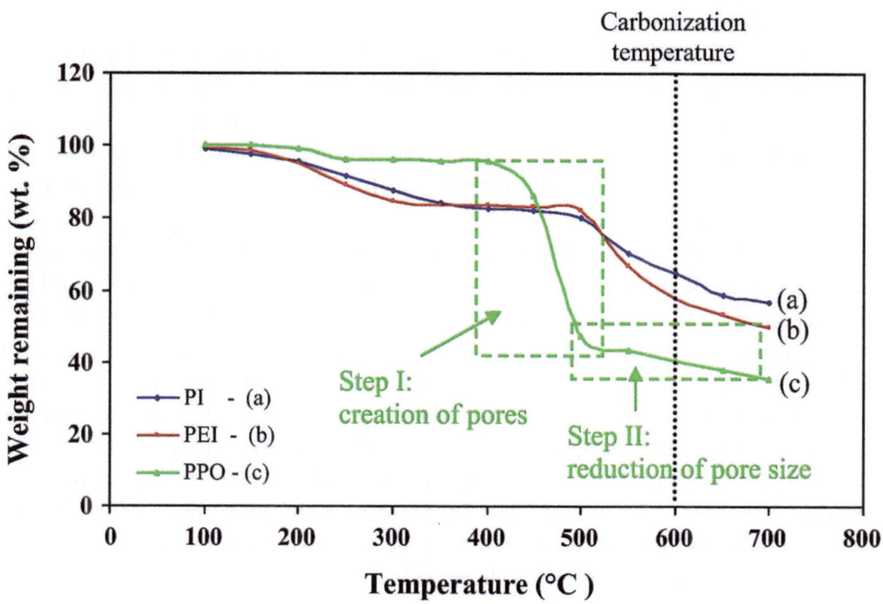


**Fig. 6** TGA curves for polymeric membranes (**a**) PI, (**b**) PEI, and (**c**) PPO [38]. Reproduced from ref. [38] with permission from Elsevier, copyright 2012

due to the thermal degradation of the benzene ring of the PEI precursor. However, at the final decomposition temperature of approximately 600–700 °C, PEI-based membranes show a slightly higher weight loss than pure PI-based membranes, which is most likely due to the breakage of ether groups in the polymer chains during the degradation process.

As shown in Fig. 6c, the thermal degradation of PPO polymer is the most stable before 430 °C, after which the degradation rate at approximately 430–500 °C drops rapidly. The polymer structure starts to degrade and develops a carbonized structure at around 430–500 °C; in addition, the weight loss of PPO at 500 °C is 53 wt%. Compared with the PI and PEI polymers, PPO shows a major weight loss in the temperature range of 400–500 °C, indicating that PPO has low thermal resistance properties. When PPO experiences huge weight loss at the temperature range of 400–550 °C, this causes a cleavage of the main group of the PPO precursor and leads to the creation of pores. Meanwhile, in the temperature range of 500–700 °C, the lower weight loss is caused by the evolution of residue gas, which leads to an incremental densification of the porous CMS membrane. These observations indicate that after carbonization at 600 °C, the PPO coating layer becomes a dense carbon structure. The remaining weight of the PPO shows the values of 40 wt% at a temperature higher than 600 °C, which is lower than those of the PEI and PI polymers. Based on these results, it can be deduced that the thickness of the PPO layer would become even thinner than those of the PI or PEI layers [38].

The gas permeances of the pristine (Pure carbon membranes) and modified PI- and PEI-derived CMS membranes were evaluated as a function of the kinetic diameters of the gas molecules. All gas permeation rates through the CMS membranes were correlated with the kinetic diameter of the gas molecules in the following order: $H_2 > CO_2 > N_2 > CH_4$ as seen in Table 5. The gas permeances obtained from the pristine PI-derived CMS membrane abruptly decrease with increasing kinetic gas diameters of 565, 168, 32.9, and 63.8 Barrier for $H_2$, $CO_2$, $N_4$ and $CH_4$, respectively. This phenomenon suggests that the permeation is controlled by the molecular sieving mechanism rather than a Knudsen diffusion mechanism for the resulting carbon membranes [38].

Compared with pristine CMS membranes, carbon segment modification increases the permeances of smaller gases, $H_2$ and $CO_2$. Moreover, the permeabilities of $H_2$ is enhanced from 565.0 and 255.0 Barrer to 1448.3 and 811.6 Barrer for the modified PI- and PEI-derived CMS membranes, respectively. Meanwhile, the permeabilities of larger gases $N_2$ and $CH_4$ decrease. The extent of the enhancement on gas permeance through the modified carbon membrane varies, depending on the different molecular sizes of the gas species. This means that the permeances of the gas species with a smaller molecular size (e.g., $H_2$ and $CO_2$) significantly increase compared with the results for $N_2$ and $CH_4$. The increase in the permeances of $H_2$ and $CO_2$ is assumed to be related to the shift of the pore size distribution to lower values and that of the pore volume to higher values [38].

A significant increase in the permselectivities of all gas pairs is observed for the modified CMS membrane compared with the results of the pristine CMS membranes (Table 5). The permselectivity of the $H_2/N_2$ of the PPO/PI shows a value of

**Table 5** Gas permeability and permselectivity measured at 298 K for modified and unmodified CMS membranes [38]

| Sample | Permeability | | | |
|---|---|---|---|---|
| | $H_2$ | $CO_2$ | $N_2$ | $CH_4$ |
| *PPO*[a] | 140.1 ± 2.7 | 45.4 ± 1.3 | 4 ± 0.01 | 4 ± 0.01 |
| *PI*[b] | 565 ± 10.5 | 168 ± 5.3 | 32.9 ± 1.2 | 63.8 ± 2.6 |
| PPO/PI | 1448.3 ± 56.9 | 1318.2 ± 37.6 | 8.4 ± 0.04 | 8.4 ± 0.03 |
| *PEI*[b] | 255 ± 2.9 | 200 ± 3 | 13.9 ± 0.2 | 10.3 ± 0.1 |
| PPO/PEI | 811.6 ± 22.7 | 512.5 ± 15.1 | 6 ± 0.02 | 6 ± 0.02 |
| **Sample** | **Permselectivity** | | | |
| | $H_2/N_2$ | $H_2/CH_4$ | $CO_2/N_2$ | $CO_2/CH_4$ |
| *PPO*[a] | 35.3 ± 0.71 | 35.3 ± 0.73 | 11.4 ± 0.29 | 11.5 ± 0.29 |
| *PI*[b] | 17.1 ± 0.69 | 8.8 ± 0.36 | 5.1 ± 0.02 | 2.6 ± 0.02 |
| PPO/PI | 171.9 ± 7.74 | 172.4 ± 7.37 | 156.4 ± 5.34 | 156.9 ± 5.03 |
| *PEI*[b] | 18.3 ± 0.13 | 24.7 ± 0.14 | 14.4 ± 0.02 | 19.4 ± 0.01 |
| PPO/PEI | 136.2 ± 4.35 | 136.3 ± 4.31 | 86 ± 2.85 | 86.1 ± 2.76 |

[a] Adapted from reference [39]
[b] Adapted from reference [40]

171.9, which is much higher than that of the PI-derived carbon membrane ($H_2/N_2 = 17.1$). The notably increased values for $H_2/CH_4$, $CO_2/N_2$, and $CO_2/CH_4$ are also shown in Table 5. The pore size for gas diffusion is narrowed down by carbon segment deposition and the formation of a barrier against the transport of larger gases $N_2$ and $CH_4$. Thus, the penetration of the larger $N_2$ and $CH_4$ molecules across the modified PPO-derived carbon layer is inhibited, while the separation ability is enhanced [38].

### *4.2 Compositional Modification Effects on Carbon Based Membranes*

Unlike its mechanical counterpart, compositional modifications on carbon based membranes are done by adding extra substances before or after fabrication that will bond them with a selected carbon membrane. This is generally done to achieve specific functions on the membrane such that it may be applicable for certain purposes, such as water treatment. This subchapter will discuss the general idea of compositional modifications on carbon based membranes and further elaborate on a selected topics of modification.

As has been discussed in previous chapters, the physical mechanics of molecular diffusion through membranes depends on porosities and membrane surfaces to promote adsorption of molecules. Terms that become familiar in membrane surface structure are hydrophobicity and hydrophilicity, the capability for an evaluated membrane to reject or attract water absorption, for membranes that are designed to interact with substances containing water. In laboratory settings, these properties measured off of diffusion are tested in specified conditions such as using preselected gasses or liquids that are very minimally polluted with non-tested substances. However, this may become an issue with practical applications of the membranes as it will be exposed to substances with pollutants of varying molecules regardless of prior filtration before becoming in contact with the membrane. These pollutants will affect the membrane's overall performance via fouling [41] and may even degrade its capabilities over time via aging [42]. As such, another important element in membranes is its permeability and selectivity.

Creating a membrane over a desired permeability and selectivity follows a specific design of porosity and membrane surface texture. This is done to make the diffusion process apply only on specific molecules in gasses as efficiently as possible promoting separation and sieving of gasses or other liquids. Compositional modification is another set of methods done to achieve further specific selectivity and diffusion of a membrane. The general idea of this method is by binding the carbon that makes up the membrane with other molecules such that the chemical binding may modify the pore sizes thus allowing only molecules of specific sizes to pass through the membrane. There are many varying methods of molecular modification of a known carbon membrane, most focusing on either molecular modification in the precursors pre-fabrication or with growth of other membranes post fabrication. This subchapter will discuss efforts to modify carbon membranes through zeolite growth [43] and silicon doping [44].

## Carbon: Zeolite Composite Membranes

Consider first membranes made out of different materials other than carbon. While they are evaluated out of the same parameters as their carbon counterparts, certain aspects of its chemical composition or fabrication techniques will leave a variety of differences and weaknesses in its capabilities, becoming one of the motivations for developing variations of substances used to produce membranes for diverse sets of application. With these variations, comes also varying amounts of membrane fabrication techniques. If these fabrication techniques of certain membranes do not affect others, they may be combined by encasing the surface of one membrane with another, in hopes of applying the benefit of surface interactions and diffusion of both membranes. One of the materials that has been used for membranes or as a supplement to other materials are zeolites. Zeolite can be described as a kind of crystalline hydrated aluminosilicate and has been used in materials and membrane studies since the 1990's [44]. The usage of zeolite as support in developing membranes has been done on ceramics previously in 2003 [45] shows great promise of performance, but difficulties in fabrication in part of costs and aluminium interaction has ostensibly made it harder to expand on quality control and hopes of applications. By applying zeolite seeds during the fabrication process of carbon membranes, the intention is for the produced membrane to have a specific and controllably uniform pore-sizes and for the final product to have an increased desired properties from the contribution and interaction of both carbon and zeolite.

**Fig. 7** SEM images of zeolite NaA seeded carbon tubes. The first picture depicts the zeolite seeded on the carbon tube's surface. The second picture depicts zeolite NaA growth post culturing. The last picture depicts the cross section of zeolite NaA seeded carbon tube with the top and middle layer showing the contribution of the zeolite [43]. Reproduced from ref. [43] with permission from Elsevier, copyright 2006

In their work, Zhang et al., 2005 (see Fig. 7) developed zeolite NaA seeds consisting of $0.33Na_2O : 3.2SiO_2 : 1Al_2O_3 : 4.0(TMA)_2O : 245H_2O$ of sizes ~150 nm prepared via clear synthesis solution to be placed within the inner surfaces of an already prepared carbon tubes of pore sizing in 0.5 μm. With the implanted seeds, the tube is then put within a clear synthesis solution of molar composition $1Al_2O_3 : 5SiO_2 : 48Na_2O : 4500H_2O$ to grow the seeds into zeolite NaA films forming within the membrane before being put in an oven at 428 K for 24 h. The final result achieved was then analyzed using XRD for zeolite formation characterization and SEM for zeolite morphology.

As visible from Fig. 8, the zeolite NaA seeds are embedded via slip casting on the exterior of the carbon tubes forming a continuous, uniform, and smooth surface thus improving on the surface structure and smoothness of the membrane. Post-hydrothermal synthesis, the zeolite membrane forms with a high uniformity and intergrowth with a noted specification of synthesis conditions may modify the desired pore sizes of the zeolite membrane. Following the completion of fabrication, permeability tests is done on the membrane using pure gasses $N_2$, $H_2$, and $C_3H_8$. Base tests done using $N_2$ returns data that showed that the selected zeolite NaA seed and synthesis made no important defects and by permeability comparison with other pure gasses shows improvement on pore sizes. The latter Fig. 8 provides additional data that provides the argument for an improved sieving ability of this composite membrane for filtering $H_2/N_2$ and $H_2/C_3H_8$ showing the promising ideal separation numbers of 24.7 and 32.2 respectively [43].

## Silicon Doped Carbon Membranes

Another method usable to compositionally modify carbon membranes can be done via the precursors pre-selected for fabrication [42]. Consider that the fabrication of carbon membranes follows a thermal process on a selected precursor to grow carbon structures with compositions mixing contributions of a majority carbon and

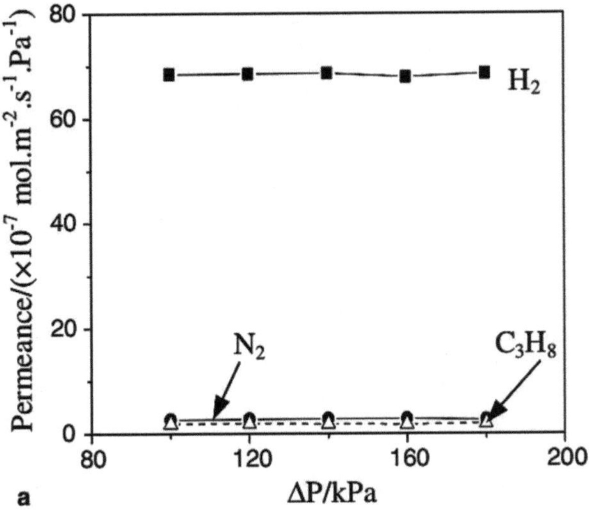


**Fig. 8** Measured permeation of selected pure gasses through the zeolite NaA seeded carbon tubes under the operating temperature 293 K [43]. Reproduced from ref. [43] with permission from Elsevier, copyright 2006

minor elements from the precursors. The idea raised in this method is to grow inherent composition throughout the membrane, for an overall uniform desired capability. For Carbon Molecular Sieving Membranes or CMSMs, these may include copolymerized precursors, blending of precursors with other polymers, or fillers have commonly been adapted to improve permeability and separation. This group's work focuses on molecular doping, and specifically using the element silicon. Molecular doping was a method developed to increase permeability, selectivity, or the aging of produced membranes though early publications revealed difficulty in improving all three aspects, most resulting in one of the characteristics being sacrificed. While early publications uses metal ions as doping, silicon has notably shown promise as it has promising thermal stability and can form covalent bonds with carbon matrix, thus possibly affecting pore formation, structural stability, and further promotes adsorption and diffusion.

This work further expands on a pre-existing work, now adding siloxane with single Si—O—Si via synthesis onto 6FDA-DABA copolymer meant to be a precursor for further CMSM production. Their method of production began by synthesis of poly(siloxane imide) or PSI onto the DABA polymer before further processing as shown in Fig. 9 which will act as the precursor. The precursor content is varied according to the molar ratio between DABA and siloxane component of the 6FDA-DABA precursor, with the noted ratios of 10:0, 9:1, 8:2, and 7:3. They will be labeled according to the PSI content within the mixture such that they are accordingly named PSI-0, PSI-1, PSI-2, and PSI-3. The precursor is prepared by soaking the PSI under NMP before soaking in methanol and then placed within an oven for pyrolysis. The final product is then tested for its permeability and selectivity, and characterized using spectroscopy akin to previous methods. Their results does a

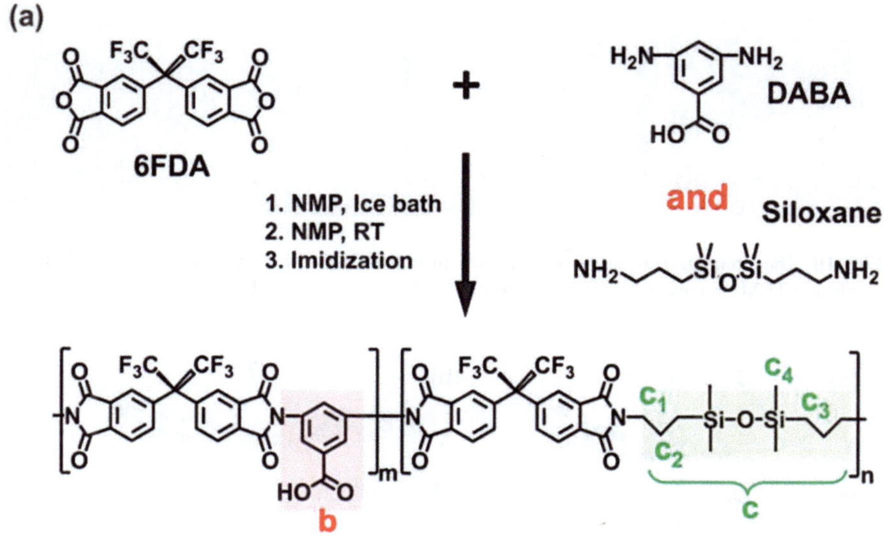


**Fig. 9** Chemical composition of the precursor PSI synthesized for making CMSM via pyrolysis [42]. Reproduced from ref. [42] with permission from Elsevier, copyright 2023

comparison between the 4 created membranes they named C-PSI, differing between molar ratios in the precursor and named correspondingly C-PSI-0, C-PSI-1, C-PSI-2, and C-PSI-3 as shown in Fig. 10.

Following the TEM characterizations in Fig. 10, it is known that the carbon structures are amorphous which implies that the Si doping has fully embedded and becomes part of the chemical bonding structure of the carbons by bonding directly through carbon or with the help of oxygen elements. For a more precise molecular binding, the authors also added the simulated model in Fig. 11 which is based on spectral data of the produced membranes. It can also be noted from the TEM images that surfaces of the carbon membrane become more ordered and thus affecting adsorptions. Specifically in Fig. 11d, it can be seen that carbon strips or graphitic sheets started to form [42]. Lastly Fig. 12 notes the permeability and selectivity of C-PSI, noting that permeability of pure gasses tends to peak at molecular ratio 8:2 or C-PSI-2, but selectivity becomes better at higher silicon content of 7:3 or C-PSI-3. Note also there is a reduced performance if the gas applied is a mixture (for example, $H_2N_2$). This is inline with the argument presented earlier in this chapter that it tends to be difficult to increase both permeability and selectivity at the same time, though it is important to note that the $H_2$ permeability of C-PSI is quite higher than several known membranes such as cellulose, phenolic resin, PPESK, PEI, and PI [42].

From both of the methods discussed in this subchapter, it is summisable that compositional modification aims to minimally modify a carbon membrane by introducing new elements into the chemical composition. The goal is to increase the effectiveness of carbon membranes by increasing uniformity and effectiveness of the contact surface of carbon membranes via pre or post fabrication chemical processes.

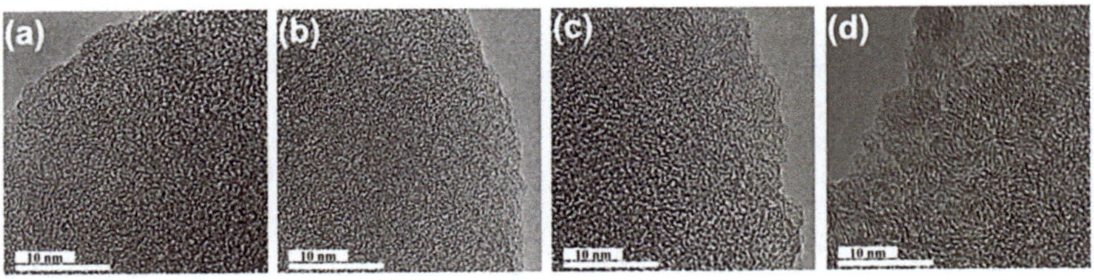


**Fig. 10** The characterization of carbon structure for CMSMs carbonized at 600 °C. High-resolution TEM images of (**a**) C-PSI-0, (**b**) C–PSI-1, (**c**) C–PSI-2 and (**d**) C–PSI-3

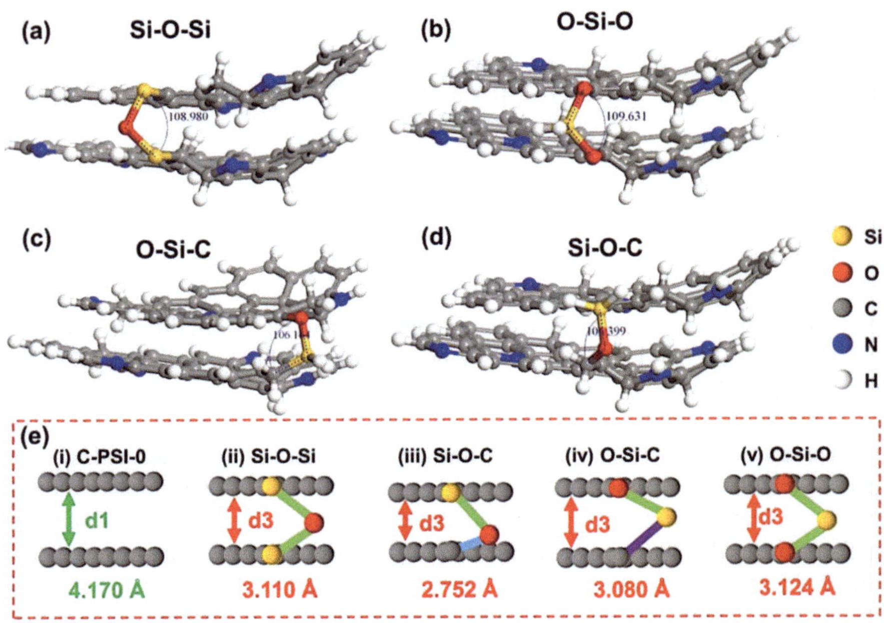


**Fig. 11** Simulation of possible Si bonding within the C-PSI membranes ordered (**a**) Si—O—Si, (**b**) O—Si—O, (**c**) O—Si—C, and (**d**) Si—O—C also (**e**) the depicted distance of carbon layers or ultra micropores of C-PSI [42]. Reproduced from ref. [42] with permission from Elsevier, copyright 2023

# 5 Theoretical Modelling and Numerical Approach to Transports on Carbon Membranes

Recurring during discussion of functionalities of any membranes are its capability to do permeation of specific gasses or liquids through its system and the selectivity of specific molecules out of mixtures to permeate. Testing of this characteristic is done experimentally using pure or selected mixtures under ideal thermal operating

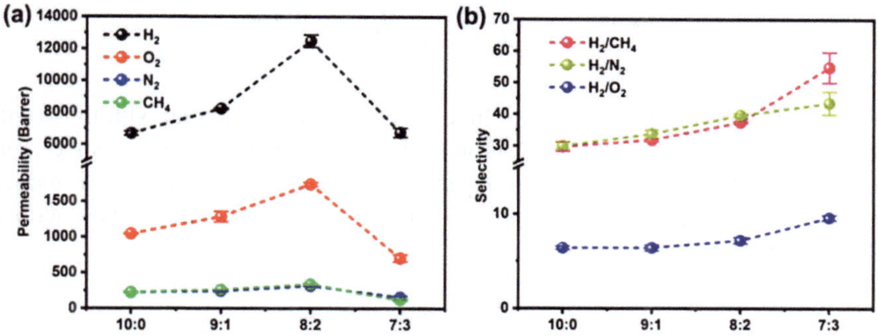


**Fig. 12** (**a**) The single gas permeability of CMSMs carbonized at 600°C under argon purge at the operating temperature of 35°C and pressure of 55 psi; (**b**) The ideal $H_2/CH_4$, $H_2/N_2$, and $H_2/O_2$ selectivity of CMSMs from the single gas permeation test

conditions of the membrane and measuring the flow rates pre and post membrane interaction. With membrane physics, physical characteristics of the membrane such as pore sizes and molecular interaction can limit what mechanics dominate molecular movement through the membrane as seen in Table 1. As such, while all diffusion or flow rates approaches to transports in membranes discuss the same parameters, certain terms will dominate more depending on the physical characterization, changing the mathematical analysis of the results. This is another motivation to do membrane characterization post fabrication, as by doing so will attain the scientist measured parametric values that will help in doing mathematical analysis of the membrane which will help with achieving consistent and efficient production and to do comparisons of experimental and control membranes of similar types.

This subchapter will discuss several of the more successful models in describing transports of molecules in carbon membranes. It is important to note that the many competing models currently being developed arise from an inherent difficulty in developing a uniform model of mass transfer in carbon membranes. This is caused by difficulty in maintaining a consistent reproducibility of carbon membranes and the complexities of the many contributing degrees of freedom [46]. Still, for carbon membranes the more successful transport models are given by the solution-diffusion model, Fick's first law, and the parallel resistance model simulated through molecular dynamics and Monte Carlo as described mathematically in Sect. 3.

## *5.1 Solution-Diffusion Model*

The solution-diffusion model is a uniform approach used in membrane mechanics, which observes the changes in chemical and thermal properties of molecules as it passes and interacts with the membranes [47]. The flux of the flow $J_i$ for a molecule of species $i$ can be generically introduced as

$$J_i = -c_i U_i \frac{d\mu_i}{dx}$$

with $U_i$ the proportional coefficient linking the chemical potential with the driving force and $c_i$ the molecular concentration, which is introduced in the derivative $d\mu_i/dx$ the gradient of chemical potential along the thickness of the membrane. This reduces most of the driving force as components of the chemical potential, where with all the limitations can be expressed as

$$d\mu_i = RT\, d\left(\ln \gamma_i n_i\right) + v_i dp$$

factoring the mole fraction $n_i$, concentration-activity coefficient $\gamma_i$, and the molar volume $v_i$ alongside its pressure. Note that the chemical potential can be derived for both compressible and non-compressible phases of the molecules interacting with the membrane. From here, is where many of the expressions of transport differ as many assumptions or contributing factors need to be introduced, especially in the factor of coefficients.

There are two approaches applying the mathematical model of solution-diffusion, given by pore-flow model with the more gradual change in pressure and constant solvent concentration, differing from the solution-diffusion model with the abrupt change in pressure and non-constant solvent concentration, as observed in Fig. 13. Both approaches however begin their mathematical analysis by evaluating the chemical potential at the contact term of liquids and the membrane. Notably, Wijmans and Baker, 2006 has done the approach of treating gas separations via membrane interaction by evaluating such that the chemical potential of the gasses

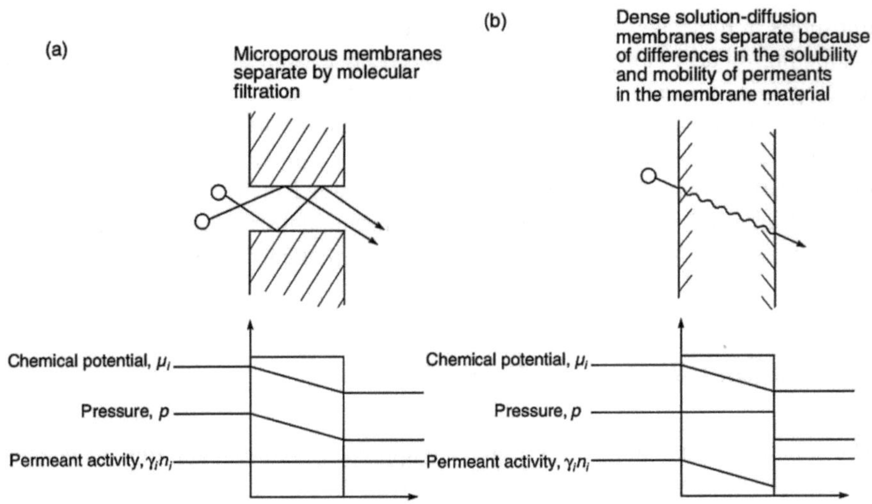


**Fig. 13** The two models derived from the solution-diffusion model, consecutively the pore-flow approach (**a**) and the solution-diffusion approach (**b**) [47]. Reproduced from ref. [47] with permission from John Wiley and Sons, copyright 2006

must be equal both internally and externally at both the feed and downstream interface, such that

$$\mu_i = \mu_{i(m)}$$

and finding the solution for the mole fraction within the membrane, which then is used to calculate the concentration and then used on the derivative of the chemical potential known as the expression of Fick's Law or the gradient of the chemical potential at the feed term (membrane density = 0) and the downstream term (membrane density = L, 1D) (Fig. 14).

## 5.2 Knudsen Diffusion

The knudsen diffusion follows a similar premise of a flux of molecules passing through a membrane [19] Beginning with the Fick's diffusion equation that is derived from the chemical potential, however now with a well defined coefficient $D_i$ such that

$$J_i = -D_i \frac{dc_i}{dx};\ D_i = D_{(\text{diffusion type}),c} \frac{d(\ln p)}{d(\ln c)}$$

this follows the pore-flow model that better fits the mechanics of carbon membranes with well defined pore diameter sizes and interaction energy of the molecules passing through with the membrane walls contributing in the coefficient and concentration terms [47]. Similarly as before, the results of these equation is for the value of fluxes of molecules that pass through the membrane which will be used to

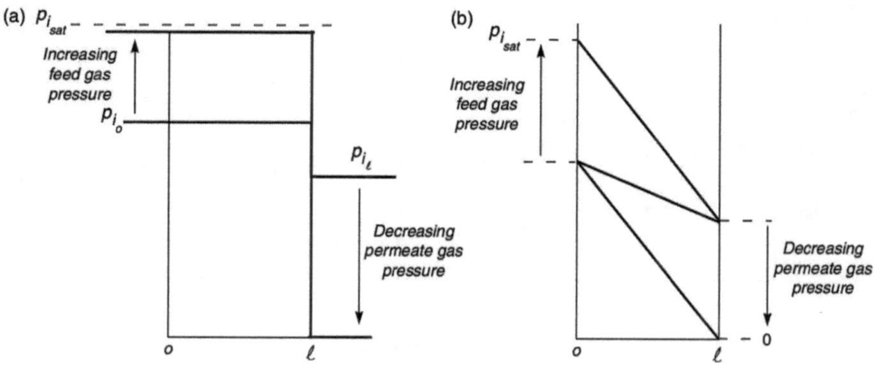


**Fig. 14** The changes calculated in pressure (**a**) and concentration (**b**) for gas separation using the solution-diffusion model [47]. Reproduced from ref. [47] with permission from John Wiley and Sons, copyright 2006

determine the permeance $P_i$ and permeability $Q_i$ which is related one another and with the flux through

$$P_i = \frac{Q_i}{L} = \frac{J_i}{\Delta p}$$

The calculations done here can be extended as shown by Gilron and Soffer in 2002 by treating the pole structures and the fluxes flowing through as akin to the resistance of particles flowing through a medium, introducing two assumptions of membrane models. The first model is treating a membrane of many but uniform pore sizes which may be approached using parallel resistance $P_{i,\text{ total}} = \sum_{\text{all diffusion}} P_{i,\text{ (model)}}$. The second model treats a single possible membrane pore as to have varying diameters along the thickness of the membrane such that the resistance model is in series resulting in $\frac{1}{P_{i,\text{total}}} = \sum_{\text{all diffusion}} \frac{1}{P_{i,(\text{model})}}$. The acquired model is then usable to predict the permeance of varying gasses passing through a selected membrane, which an example is graphed in Fig. 15 The selectivity of molecules $\alpha_{ij}$ with species i and j is simply a ratio of permeability between the molecules that is desired to pass through the membrane with other molecules with the mixture of gas or liquid that interacts with the membrane, commonly approached using

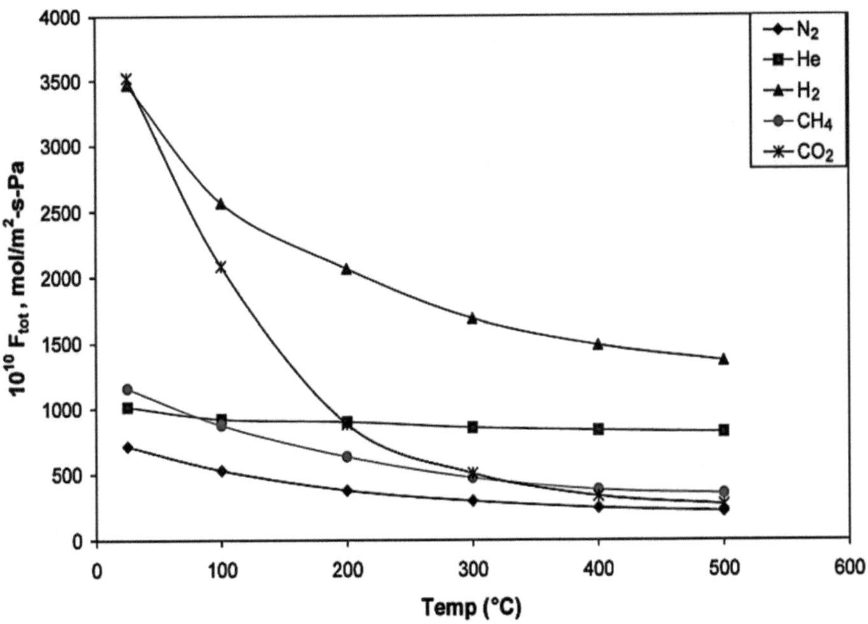


**Fig. 15** The plotted permeance of various gasses as a function of temperature on a Hollow Fiber Carbon Membrane of a selected length measurement [48]. Reproduced from ref. [47] with permission from Elsevier, copyright 2022

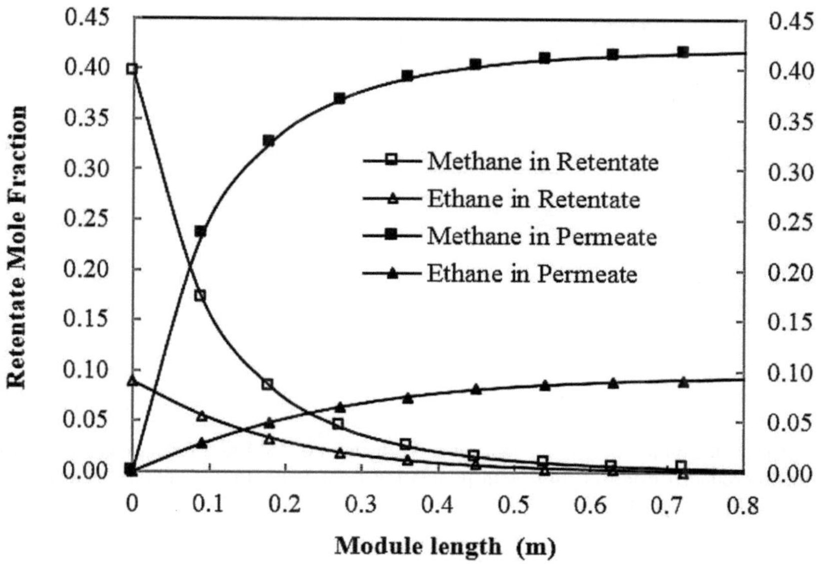


**Fig. 16** Permeance mole fraction of Methane and Ethane in CMSM as a function of membrane length [49]. Reproduced from ref. [49] with permission from Journal of Membrane Science and Research (JMSR), copyright 2014

$$\alpha_{\text{pass,reject}} \approx \frac{P_{\text{pass}}}{P_{\text{reject}}}$$

Further progress in mathematical modeling of diffusion in carbon membranes remains consistent with the principles derived in this subchapter. The models presented in Sects. 3 and 4 follow this framework while incorporating the dependence of the diffusion coefficient on physical and thermodynamic parameters such as pore size, pressure, and molecular interaction energies. Figs. 16, 17, 18 and 19 (from Gholami et al. [48]) illustrate the independent effects of membrane length (Fig. 16), membrane area (Fig. 17), feed pressure (Fig. 18), flow rate (Fig. 19), and temperature (Fig. 20) on the permeance of a methane ($CH_4$) and ethane ($C_2H_6$) mixture using a carbon molecular sieve membrane (CMSM). The results were validated with the data summarized in Tables 6 and 7, following the mathematical model described in Sect. 4.

From the results derived by Gholami et al. [48], the relations and individual contributions of physical and thermodynamic parameters on the permeance can be shown such that an observer should be able to determine the parametric values that makes a carbon membrane of great performance. As visible in Figs. 16 and 17, the greatest contributing characteristics that affect transports in carbon membranes are the length and area dimension.

It is deductible that the mathematical modelling of transport in membranes is done via diffusion mechanics to describe the fluxes of molecular mechanics

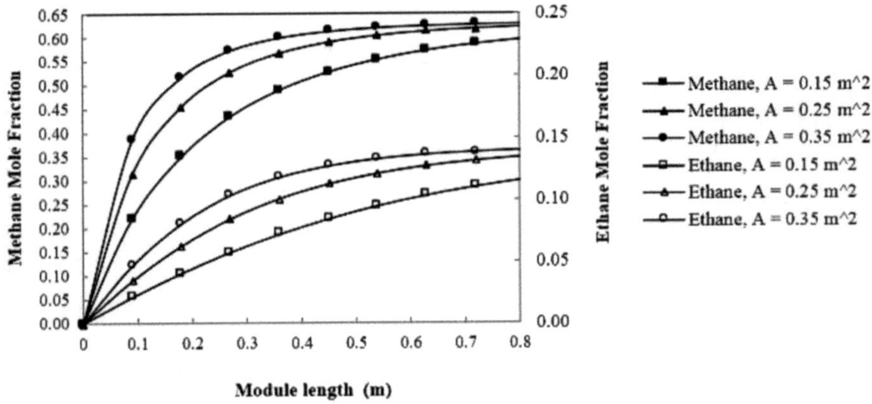


**Fig. 17** Permeance mole fraction of Methane and Ethane in CMSM as a function of membrane area [49]. Reproduced from ref. [49] with permission from Journal of Membrane Science and Research (JMSR), copyright 2014

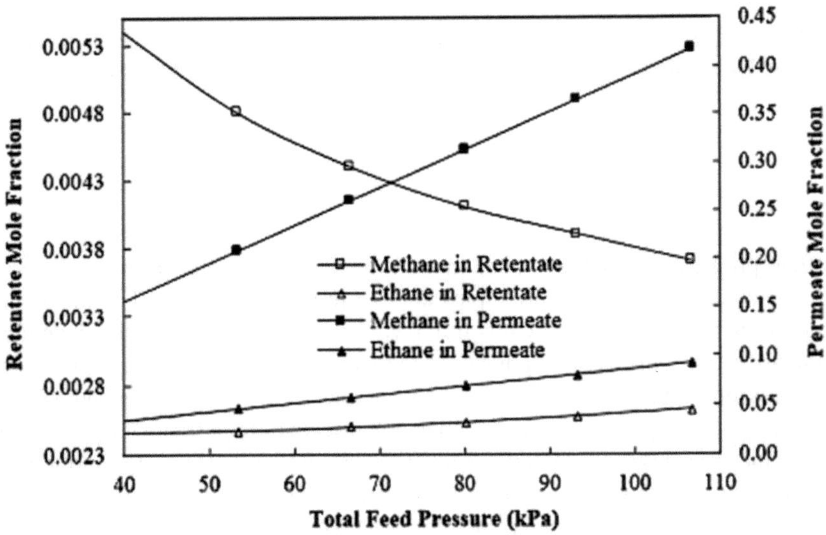


**Fig. 18** Permeance mole fraction of Methane and Ethane in CMSM as a function of feed pressure [49]. Reproduced from ref. [49] with permission from Journal of Membrane Science and Research (JMSR), copyright 2014

through the membrane to find the related quantity permeance and selectivity. The numerical analysis done on the equation strives to find how parameters of both the interacting liquids and the physical characteristics of the membrane contribute to the final permeance. This is done by fitting curves and mathematical expression onto known data of membranes, and varying certain parameters individually as simplification.

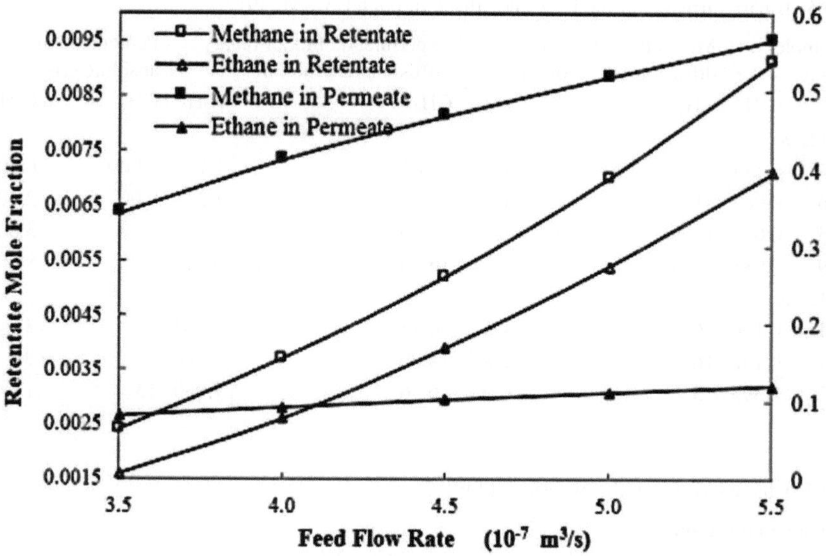


**Fig. 19** Permeance mole fraction of Methane and Ethane in CMSM as a function of feed flow rate [49]. Reproduced from ref. [49] with permission from Journal of Membrane Science and Research (JMSR), copyright 2014

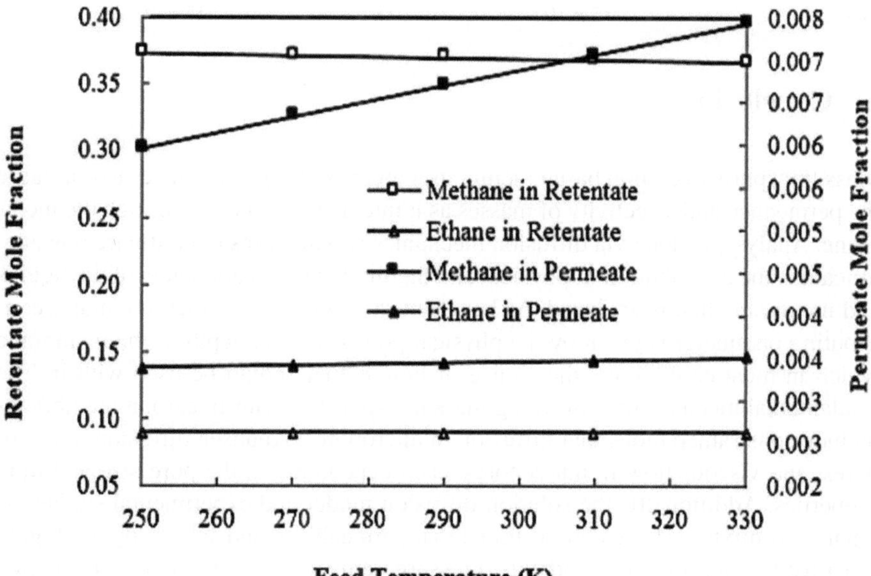


**Fig. 20** Permeance mole fraction of Methane and Ethane in CMSM as a function of feed temperature [49]. Reproduced from ref. [49] with permission from Journal of Membrane Science and Research (JMSR), copyright 2014

**Table 6** Binary diffusion of methane and ethane through CMSM [49]

| Feed mole fraction | | Model flux of binary diffusion $10^{-5}$ mol. $m^{-2}$. $s^{-1}$ | | Experimental flux of binary diffusion $10^{-5}$ mol. $m^{-2}$. $s^{-1}$ | | Percentage of absolute error | |
|---|---|---|---|---|---|---|---|
| $CH_4$ | $C_2H_6$ | $CH_4$ | $C_2H_6$ | $CH_4$ | $C_2H_6$ | $CH_4$ | $C_2H_6$ |
| $T = 323$ K | | | | | | | |
| 0.203 | 0.117 | 4.80 | 1.62 | 5.63 | 1.73 | 14.00 | 6.00 |
| 0.316 | 0.062 | 1.20 | 1.20 | 1.10 | 1.23 | 9.00 | 2.40 |
| 0.395 | 0.089 | 12.70 | 1.28 | 12.80 | 1.32 | 0.78 | 3.00 |
| 0.499 | 0.088 | 18.00 | 1.02 | 16.40 | 0.94 | 9.70 | 8.80 |
| $T = 353$ K | | | | | | | |
| 0.221 | 0.106 | 6.3 | 1.8 | 6.43 | 1.743 | 2 | 3 |
| 0.477 | 0.142 | 16.0 | 1.8 | 14.00 | 1.990 | 14 | 9 |
| 0.252 | 0.395 | 5.7 | 6.9 | 6.56 | 7.950 | 13 | 13 |

**Table 7** Characteristics of the experiment [49]

| **Transport characteristics** | | | |
|---|---|---|---|
| $CH_4$ permeance ($m^3$. $m^{-2}Pa^{-1}$. $s^{-1}$) | | $6 \times 10^{-7}$ | |
| $C_2H_6$ Permeance ($m^3$. $m^{-2}Pa^{-1}$. $s^{-1}$) | | $2 \times 10^{-7}$ | |
| Selectivity $\left(\frac{CH_4}{C_2H_6}\right)$ | | 2.73 | |
| **Characteristics of the Carbon Membrane** | | | |
| **Effective area ($m^2$)** | **Thickness (m)** | **Length (m)** | **Density (kg $m^{-3}$)** |
| 0.3 | $15 \times 10^{-6}$ | 0.8 | $2.15 \times 10^3$ |

# 6 Conclusion

Mass transports in carbon based membranes are modelling approaches to configure the permeance and selectivity of masses as it interacts and moves through the membrane. Analysis is done via diffusion mechanics and attempts to configure how significant is the contribution of physical and thermodynamic parameters of the liquids and membrane that is analyzed. As has been shown in this chapter, the major contributing parameter is given by the physical pore sizes and depth of the membrane which in most cases limits the choice of liquids that should be used with it. The fundamental mechanisms governing mass transport behavior in carbon membranes include constrained molecular diffusion in micropores, Knudsen diffusion in mesopores, and viscous flow in macropores, each depending on the pore size and fluid properties. Additionally, the solution-diffusion model and experimental studies on binary gas mixtures have demonstrated that permeability and selectivity are highly sensitive to concentration gradients, pressure, temperature, and molecular interactions with the membrane structure. However, this arises an inherent difficulty of keeping consistency and homogeneity of the created carbon membranes caused by the fabrication process. While mathematical modelling of mass transports yields the ideal parametric values to create for in the membrane, any produced carbon membrane tends to be anisotropic with chaotic surfaces and varying thickness in the pore walls which would yield experimental data that varies in a manner that can not be ignored. As such, future experiments on carbon based membranes for effective mass transport should first increase the uniformity of desired physical values within fabricated membranes and to produce them as consistently as possible. Recent advances in theoretical modeling and numerical simulations have provided deeper insights into the fundamental transport phenomena, revealing the complex interplay between pore morphology, molecular size, and operational conditions. Recent attempts at solving this problem involve modification of the precursor material, committing to highly specific fabrication processes, or coating the carbon membrane with other membranes in which all proposals have shown great promise.

## References


1. Widakdo, J., et al.: Graphene oxide-melamine nanofilm composite membrane for efficient $CO_2$ gas separation. Sep. Purif. Technol. **323**, 124521 (2023)
2. Widakdo, J., et al.: Bioinspired ionic liquid-graphene based smart membranes with electrical tunable channels for gas separation. Appl. Mater. Today. **27**, 101441 (2022)
3. Widakdo, J., et al.: Mechanism of a self-assembling smart and electrically responsive PVDF–graphene membrane for controlled gas separation. ACS Appl. Mater. Interfaces. **12**(27), 30915–30924 (2020)
4. Widakdo, J., et al.: Switching gas permeation through smart membranes by external stimuli: a review. J. Mater. Chem. A. **10**(32), 16743–16760 (2022)
5. Widakdo, J., et al.: Tailoring of graphene–organic frameworks membrane to enable reversed electrical-switchable permselectivity in $CO_2$ separation. Carbon. **182**, 545–558 (2021)
6. Austria, H.F.M., et al.: A review on the recent advancements in graphene-based membranes and their applications as stimuli-responsive separation materials. J. Mater. Chem. A. **9**(38), 21510–21531 (2021)
7. Widakdo, J., et al.: Positron annihilation spectroscopy for the free volume depth profile analysis of multilayer and 2D materials composite membranes: a review. Sep. Purif. Technol. **322**, 124366 (2023)
8. Abebe, S., et al.: Zwitterionic M-Pha complex modified lamellar go nanocomposite membranes for enhanced organic solvent nanofiltration with antifouling performance. J. Membr. Sci. **736**, 124596 (2025)
9. Foorginezhad, S., et al.: Emerging membrane technologies for sustainable water treatment: a review of recent advances. Environ. Sci.: Adv. **4**, 530–570 (2025)
10. Austria, H.F.M., et al.: Investigation of the pH-mediated fabrication process of pure and polyethyleneimine-crosslinked graphene oxide membranes for desalination and heavy metal ion separation. Carbon. **224**, 119019 (2024)
11. Austria, H.F.M., et al.: Tailoring the specific crosslinking sites of graphene oxide framework nanosheets for controlled nanofiltration of salts and dyes. J. Clean. Prod. **395**, 136280 (2023)
12. Subrahmanya, T., et al.: Effect of functionalized nanodiamonds and surfactants mediation on the nanofiltration performance of polyamide thin-film nanocomposite membranes. Desalination. **555**, 116540 (2023)
13. Abebe, S.H., et al.: Zwitterionic M-PhA modified GO nanocomposite membranes for enhanced organic solvent nanofiltration and wastewater treatment with antifouling performance. J. Membr. Sci., 124596 (2025)

14. Sun, P., Wang, K., Zhu, H.: Recent developments in graphene-based membranes: structure, mass-transport mechanism and potential applications. Adv. Mater. **28**(12), 2287–2310 (2016)
15. Verweij, H.: Ceramic membranes: morphology and transport. J. Mater. Sci. **38**(23), 4677–4695 (2003)
16. Chen, X.-C., et al.: Engineering polymeric nanofluidic membranes for efficient ionic transport: biomimetic design, material construction, and advanced functionalities. ACS Nano. **16**(11), 17613–17640 (2022)
17. Lee, J., Kim, J., Hyeon, T.: Recent progress in the synthesis of porous carbon materials. Adv. Mater. **18**(16), 2073–2094 (2006)
18. Soboleva, T., et al.: On the micro-, meso-, and macroporous structures of polymer electrolyte membrane fuel cell catalyst layers. ACS Appl. Mater. Interfaces. **2**(2), 375–384 (2010)
19. Chen, J., et al.: The effect of Knudsen diffusion and adsorption on shale transport in nanopores. Adv. Mech. Eng. **9**(9), 1687814017721858 (2017)
20. Amann-Hildenbrand, A., Ghanizadeh, A., Krooss, B.M.: Transport properties of unconventional gas systems. Mar. Pet. Geol. **31**(1), 90–99 (2012)
21. Dou, H., et al.: Microporous framework membranes for precise molecule/ion separations. Chem. Soc. Rev. **50**(2), 986–1029 (2021)
22. Erdosy, D.P., et al.: Microporous water with high gas solubilities. Nature. **608**(7924), 712–718 (2022)
23. Smit, B., Maesen, T.L.: Molecular simulations of zeolites: adsorption, diffusion, and shape selectivity. Chem. Rev. **108**(10), 4125–4184 (2008)
24. Kärger, J., Valiullin, R.: Mass transfer in mesoporous materials: the benefit of microscopic diffusion measurement. Chem. Soc. Rev. **42**(9), 4172–4197 (2013)
25. Wang, H., et al.: Lattice Boltzmann model for oil/water two-phase flow in nanoporous media considering heterogeneous viscosity, liquid/solid, and liquid/liquid slip. SPE J. **27**(06), 3508–3524 (2022)
26. Dullien, F.A.: Porous Media: Fluid Transport and Pore Structure. Academic Press (2012)
27. Hebbar, R.S., et al.: Carbon nanotube-and graphene-based advanced membrane materials for desalination. Environ. Chem. Lett. **15**(4), 643–671 (2017)
28. Verweij, H., Schillo, M.C., Li, J.: Fast mass transport through carbon nanotube membranes. Small. **3**(12), 1996–2004 (2007)
29. Kärger, J., Ruthven, D.M.: Diffusion in nanoporous materials: fundamental principles, insights and challenges. New J. Chem. **40**(5), 4027–4048 (2016)
30. Roggendorf, H., et al.: Influence of temperature and water vapour pressure on drying kinetics and colloidal microstructure of dried sodium water glass. Adv. Chem. Eng. Sci. **5**(01), 72 (2014)
31. Kuila, U., Prasad, M., Kazemi, H.: Assessing Knudsen flow in gas-flow models of shale reservoirs. Can. Soc. Explor. Geophys. **38**, 23–27 (2013)
32. Kuila, U., Prasad, M.: Specific surface area and pore-size distribution in clays and shales. Geophys. Prospect. **61**(2-Rock Physics for Reservoir Exploration, Characterisation and Monitoring), 341–362 (2013)
33. Derjaguin, B., Churaev, N.: Structure of the boundary layers of liquids and its influence on the mass transfer in fine pores. In: Progress in Surface and Membrane Science, pp. 69–130. Elsevier (1981)
34. Ismail, A.F., et al.: Carbon-Based Membranes for Separation Processes. Springer (2011)
35. Liu, D.-J., et al.: Kinetic Monte Carlo simulation of statistical mechanical models and coarse-grained mesoscale descriptions of catalytic reaction–diffusion processes: 1D nanoporous and 2D surface systems. Chem. Rev. **115**(12), 5979–6050 (2015)
36. Chen, Y., Yang, R.: Preparation of carbon molecular sieve membrane and diffusion of binary mixtures in the membrane. Ind. Eng. Chem. Res. **33**(12), 3146–3153 (1994)
37. Itta, A.K., Tseng, H.-H.: Hydrogen separation performance of CMS membranes derived from the imide-functional group of two similar types of precursors. Int. J. Hydrog. Energy. **36**(14), 8645–8657 (2011)

38. Tseng, H.-H., Itta, A.K.: Modification of carbon molecular sieve membrane structure by self-assisted deposition carbon segment for gas separation. J. Membr. Sci. **389**, 223–233 (2012)
39. Gayathri, V., Devi, N., Geetha, R.: Hydrogen storage in coiled carbon nanotubes. Int. J. Hydrog. Energy. **35**(3), 1313–1320 (2010)
40. Na, Y., et al.: Stable operation of air-blowing direct methanol fuel cell stacks through uniform oxidant supply by varying fluid flow fixtures and developing the flow sensor. Int. J. Hydrog. Energy. **36**(15), 9205–9215 (2011)
41. Chen, X., et al.: The surface modification of coal-based carbon membranes by different acids. Desalin. Water Treat. **51**(28–30), 5855–5862 (2013)
42. Li, H., et al.: Simultaneously enhanced gas permeability, selectivity and aging stability of carbon molecular sieve membranes by the molecule doping of silicon. Carbon. **203**, 47–58 (2023)
43. Zhang, X., et al.: Modification of carbon membranes and preparation of carbon–zeolite composite membranes with zeolite growth. Carbon. **44**(3), 501–507 (2006)
44. Li, Y., et al.: Preparation, mechanism and applications of oriented MFI zeolite membranes: a review. Microporous Mesoporous Mater. **312**, 110790 (2021)
45. Lai, Z., et al.: Microstructural optimization of a zeolite membrane for organic vapor separation. Science. **300**(5618), 456–460 (2003)
46. Sahin, Z., et al.: Mass transport in carbon membranes. Curr. Opin. Chem. Eng. **39**, 100896 (2023)
47. Wijmans, J.G., Baker, R.W.: The solution-diffusion model: a unified approach to membrane permeation. Mater. Sci. Membr. Gas Vapor Sep. **1**, 159–189 (2006)
48. Gilron, J., Soffer, A.: Knudsen diffusion in microporous carbon membranes with molecular sieving character. J. Membr. Sci. **209**(2), 339–352 (2002)
49. Gholami, G., Soleimani, M., Takht Ravanchi, M.: Mathematical modeling of gas separation process with flat carbon membrane. J. Membr. Sci. Res. **1**(2), 90–95 (2015)

# Carbon-Based Membranes for Sustainable Energy and Environmental Applications

**Fena Kharisma Annisa, Haidah Novianti, Muhammad Daffa Nuur Hidayat, Januar Widakdo, and Aditya Rianjanu**

## 1 Introduction

Water is one of the most essential resources required by all living organisms, especially humans [1, 2]. With the continuous increase in human population growth over time, the demand for water has become significantly urgent and elevated [3]. Currently, more than 70% of the Earth's surface is covered by water, including oceans, rivers, and other bodies of water [4]. However, nearly 97% of this water is saline, leaving only about 3% as freshwater [4]. The saline water on Earth's surface contains a very high concentration of dissolved mineral salts; according to the National Oceanic and Atmospheric Administration (NOAA), approximately 3.5% by weight of seawater consists of dissolved salts [5]. Due to its high salinity, seawater cannot be consumed directly by the human body without undergoing a desalination process[6]. The development and continued research in this field remain critical, as scientists strive to advance technologies that improve desalination processes [7]. One promising approach to address this challenge is the development of nanotechnology-based membrane materials that exhibit high permeability and antifouling capabilities [8].

Carbon-based membranes have emerged as one of the most important materials in various technological applications, especially in water purification [9–12], gas

F. K. Annisa · H. Novianti · M. D. N. Hidayat · J. Widakdo
Department of Physics, Faculty of Mathematics and Natural Sciences, Universitas Indonesia, Depok, Indonesia

A. Rianjanu (✉)
Department of Materials Engineering, Institut Teknologi Sumatera, Terusan Ryacudu, Way Hui, Jati Agung, Lampung, Selatan, Indonesia

Center for Green and Sustainable Materials, Institut Teknologi Sumatera, Terusan Ryacudu, Way Hui, Jati Agung, Lampung, Selatan, Indonesia
e-mail: aditya.rianjanu@mt.itera.ac.id


H. Abdullah, J. Widakdo (eds.), *Membrane Physics of Carbon-Based Materials*, Engineering Materials, https://doi.org/10.1007/978-981-95-5605-2_4

separation [13–18], and sustainable environmental management [19–22]. The advantages of this material lie in its unique properties, such as a very high specific surface area [23], excellent chemical resistance [24], and the ability to possess functional groups that can be tailored to meet specific application requirements [25]. These characteristics allow carbon-based membranes to selectively remove various pollutants, including solid particles, ions, and even gases dissolved in liquids [26].

The growing demand for efficient and environmentally friendly treatment technologies, along with increasing concerns over resource scarcity and the need for better environmental management, has driven intensive research into more advanced membrane materials [27]. One of the most widely researched technologies today is the use of carbon-based membranes, which are expected to offer better solutions in addressing environmental challenges such as water contamination and air pollution[28].

However, despite promising results in the research and development of carbon-based membranes, many challenges remain ranging from material design and performance to economic aspects [29]. Therefore, this report aims to provide a more in-depth overview of recent innovations in carbon-based membrane design and how integration with hybrid and composite structures can enhance their performance [30]. In addition, this report will explore the role of artificial intelligence (AI) and computational modeling in optimizing membrane design, which can accelerate both research and production processes [31]. Finally, the prospects and challenges of commercializing carbon-based membrane technology will also be analyzed, with a focus on potential barriers to large-scale implementation and the steps needed to overcome them [29].

## 2 Hybrid and Composite Membrane Material Integration

If the combination of different separation methods results in a finer approach than two techniques employed individually, a hybrid or integrated process is known as '1 + 1 > 2' process [32]. Membrane integration methodology was introduced in the mid-1980s [33]. Hybrid membrane processes are those in which one or even more membrane processes are combined with other unit processes such as coagulation, adsorption, and ion exchange [34]. To complete a given task, all of these procedures are merged into a single system [35]. Hybrid materials generally combine an organic phase and an inorganic phase together at a molecular level through chemical bonds [36]. By combining the two different materials, the new hybrid membrane has novel properties such as enhanced stability, modified membrane structure and better permeation performance [37]. The interactions between the organic and inorganic components include weak hydrogen bonding and strong ionic or covalent bonding [38]. The latter is beneficial in avoiding the formation of nonselective voids at the organic–inorganic interface and agglomeration of inorganic particles [39].

Hybrid and composite membranes represent a rapidly growing area of innovation in carbon-based membrane research. Hybrid membranes combine carbon-based materials with polymers, ceramics, or other substances to enhance both performance and durability [40]. For example, polymers used in conjunction with graphene oxide or carbon nanotubes (CNTs) can improve membrane flexibility, allowing them to be applied in various environmental conditions [41]. In addition, the incorporation of metals or ceramics into carbon-based membranes enhances their thermal resistance and structural stability, making these membranes more resistant to extreme temperatures and high pressure [29].

Furthermore, research on carbon-based membranes also focuses on the development of composites that integrate carbon materials with metal-organic frameworks (MOFs)[42]. MOFs possess highly specific properties for adsorbing certain molecules, enabling these membranes to be utilized in gas separation and water treatment applications[43]. This material combination not only improves adsorption capacity but also enhances fouling resistance and increases the membrane's durability against degradation [44].

## 3 Opportunities and Current Progress in Carbon-Based Membrane Development

In the advancement of separation and purification technologies, a variety of adsorbent materials have been explored, including biosorbents, activated carbon, biochar, polymers, and nanoparticles [45]. Each comes with its own advantages and limitations in terms of efficiency, cost, and environmental impact [46]. Among them, activated carbon and other carbon-based materials such as carbon nanotubes (CNTs) and graphene offer the most promising membrane performance [47]. However, these benefits are accompanied by a number of complex challenges, particularly when it comes to large-scale commercialization (Table 1) [48].

One of the primary obstacles is the high production cost. The synthesis of CNTs and graphene requires advanced technologies and expensive raw materials [50]. Moreover, large-scale industrial production of carbon-based materials remains limited, which directly affects the high price of final products [51]. This challenge is compounded by issues of scalability, methods that are successful in the laboratory may not be feasible or cost effective when applied to mass production while maintaining consistent quality [52].

Another critical issue is membrane regeneration. While carbon-based membranes are known for their thermal stability and chemical resistance, fouling, or the accumulation of particles on the membrane surface, continues to hinder performance [29, 53]. Effective and economical regeneration methods are still under development and not yet widely available for commercial applications [54]. Environmental concerns also play a significant role. Due to their nano-scale size and lightweight nature, materials such as CNTs have the potential to be released into the environment during application, especially in water treatment processes [55]. These materials are known to be toxic to living organisms and are not easily

**Table 1** Cost and environmental impact of popular adsorbents [44, 49]

| Adsorbent type | Economic aspects | Environmental impacts | Sustainability considerations |
|---|---|---|---|
| Biosorbents | Generally low cost, varies with source material | Often environmentally benign, biodegradable | Minimizes waste, promotes resource recovery |
| Activated Carbon | Can be high, depending on production method | High energy consumption in production, chemical usage concerns | Requires sustainable production and disposal methods |
| Biochar | Variable, often low cost | Carbon sequestration, may involve high energy consumption | Utilizes waste materials, reduces greenhouse gases |
| Clays & Minerals | Cost varies with material and processing | Low to moderate environmental impact | Often involves natural, abundant materials |
| Polymers | Varies widely, often higher than natural adsorbents | Potential chemical pollution during synthesis | Needs eco-friendly synthesis and disposal strategies |
| Nanoparticles | Generally high cost | Potential ecological risks, energy-intensive production | Requires careful assessment of lifecycle impacts |
| Composites | Highly variable, dependent on components | Depends on constituent materials | Needs careful design for environmental sustainability |

biodegradable, raising concerns about their long-term impact as potential pollutants [56]. In addition, standardization and regulation remain major hurdles. For carbon-based membranes to be widely adopted in industries, particularly in water purification or gas separation, they must meet strict safety and environmental standards [26]. Unfortunately, the development of international standards for this technology is still in its early stages, slowing down its broader adoption in the market [57].

In summary, while carbon-based membranes hold tremendous promise for industrial applications, their development still faces numerous challenges that must be addressed. These include improving production efficiency, ensuring environmental sustainability, and establishing clear regulatory frameworks to support their safe and widespread use.

## *3.1 Improved Separation Performance*

One of the most promising opportunities in the development of membrane technology today lies in the use of carbon-based materials, such as graphene oxide (GO) [10–12, 14, 15, 18, 22, 58–60] and carbon nanotubes (CNTs) [61, 62], which have shown the potential to significantly enhance separation performance compared to traditional polymer-based membranes. These carbon-based materials possess outstanding physical and chemical properties, including extremely high surface area, excellent mechanical strength, good thermal stability, and tunable surface chemistry [63–65]. These characteristics directly contribute to improvements in both permeability and selectivity, the two key parameters in membrane separation processes [66, 67].

In the context of water and wastewater treatment, carbon-based membranes deliver much more efficient and selective performance[26]. For instance, a PVA-based nanocomposite film modified with GO-$NiFe_2O_4$ nanoparticles has demonstrated a high heavy metal removal efficiency ranging from 81% to 98%, even at metal concentrations of 5–20 mg/L [68]. Moreover, the incorporation of carbon materials also enhances the membrane's mechanical and thermal properties, which is crucial for long-term operational stability and durability [26]. On the other hand, graphene oxide polyamide membranes synthesized via interfacial polymerization exhibit very high water permeability exceeding 60 L/$m^2$ h, and 100% contaminant rejection at a concentration of 100 mg/L. These findings prove that carbon materials can not only significantly boost separation performance, but also maintain the quality of the final product, especially in clean water applications.

This superior performance opens up vast opportunities for industrial applications, particularly in areas such as seawater desalination, domestic and industrial wastewater treatment, gas separation, and purification processes in the chemical and pharmaceutical industries. The combination of high efficiency, tunability, and potential integration with other methods, such as photocatalysis, positions carbon-based membranes as a leading candidate for the next generation of separation technologies. If developed optimally, these materials can play a pivotal role in addressing clean water challenges and enabling more sustainable industrial processes (Table 2).

## *3.2 Innovation in Nanotechnology*

With advancements in nanotechnology, the performance and production cost efficiency of carbon-based membranes can be significantly improved [26]. Nanotechnology enables the development of carbon materials that are more efficient in filtering contaminants, more durable due to reduced fouling and degradation, and more cost-effective to produce [69]. This opens up broad opportunities for applying carbon membranes, particularly in water purification and wastewater treatment [26].

One key strategy is the surface modification of graphene oxide (GO) and carbon nanotubes (CNT) by adding functional groups such as –OH and –COOH [70, 71]. This modification enhances the hydrophilicity of the membrane surface, thereby reducing fouling and extending membrane lifespan [72]. Additionally, the integration of carbon nanomaterials with polymers or biochar leads to the creation of composite materials that are more flexible, resistant to fouling, and economically viable [64]. Another innovative approach is green synthesis, which involves producing carbon materials using organic or inorganic waste as raw materials [73]. Examples include banana peels, rice husks, corn cobs, paper waste, and coconut waste, which

**Table 2** Some recent advances in membrane based adsorption technologies for water remediation

| Membrane type | Synthesis/method | Main findings/performance |
|---|---|---|
| Multilayered Molecularly Imprinted Composite | Incorporating porous carbon nanospheres and polydopamine | 2.72× surface area increase, improved hydrophilicity, 51.40 mg/g phenol adsorption, perm-selectivity >7.6 |
| Magnetic Ion-Imprinted PAN | Polyacrylonitrile electro-spun nanofibers | 130.0 mg/g adsorption for Pb(II), selectivity coefficient of 2504 |
| MIL-53-$NH_2$ MOF/CA Membrane | In–situ incorporation of MIL-53-$NH_2$ MOF/CA | Adsorption of chlorpyrifos increased from 160.36 to 356.34 g/mg |
| PVA Nanocomposite Films | Incorporation of magnetic GO-Ni-$Fe_2O_4$ nanoparticles | Heavy metal removal efficiency 81 – 98% of 5 – 20 mg/L, improved mechanical/thermal properties |
| Poly-cyclodextrin and ZnO Nanofibrous | Electrospinning and atomic layer deposition | Surface area increase, photocatalytic activity 83.1–\94.3% (10 – 20 mg/L) |
| Graphene Oxide Polyamidoamine Membranes | Interfacial polymerization | Water permeability 35.89 – 61.52 L/ $m^2$ $h^{-1}$, 100% (100 mg/L) rejection of contaminants |
| PAA/NM88B/GO Membrane | Vacuum filtration | 98.79% (20 mg/L) removal of methylene blue, photo-Fenton catalysis |
| Poly(amide-sulfonamide) Composite Membrane | Surface activation assisted multi-step polymerization | (>99.3% of 0.1 g $L^{-1}$) colour removal in acidic wastewater treatment |
| TA-Fe Nanofiltration Membrane | Tannic acid and ferric ions coordination complexes | (>91.9% of 10 mg/L) rejection of antibiotics in saline wastewater |

can be converted into activated carbon, biochar, or carbon nanoparticles. These eco-friendly materials can be incorporated into membrane fabrication, lowering production costs while supporting sustainable waste management practices.

Furthermore, the use of photocatalytic nanoparticles such as zinc oxide (ZnO) and titanium dioxide ($TiO_2$) provides additional functional benefits [74, 75]. These nanoparticles offer antibacterial and self-cleaning properties that help prevent biological contamination and extend membrane service life. From a production standpoint, efficient manufacturing processes such as vacuum filtration, casting, and electrospinning allow for scalable membrane production using relatively simple equipment, making the technology accessible and practical. Through these various innovative approaches, nanotechnology plays a vital role in advancing carbon-based membranes that are high-performing, environmentally friendly, and cost-effective in both material and production processes [76].

## 3.3 *Current Progress in Carbon-Based Membrane Technology*

Exploration of new materials and molecular-scale modifications in membrane material design, combined with membrane fabrication, has identified a new focal point in desalination membrane technology [77]. It has been found that several carbon-based nanomaterials [78], including carbon nanotubes (CNTs) [79], carbon quantum dots (CQDs) [80], and graphene oxide (GO) [81], hold great potential as next-generation membrane technologies due to their unique properties such as rapid mass transport rates and distinctive structural characteristics [82]. By integrating existing desalination membranes with carbon-based nanomaterials, significant improvements in membrane performance for mineral salt separation processes are expected [29].

Research conducted by Hummer et al. in 2001 on the permeability behavior of CNTs demonstrated that when water flows through the relatively narrow CNT channels, water molecules can permeate by forming interconnected chain-like structures facilitated by hydrogen bonding between molecules [83]. This discovery established the substantial potential of CNTs in water separation applications, including seawater desalination.

Moreover, two-dimensional atomic carbon sheets, known as graphene, have been extensively studied and developed [84]. By stacking multiple graphene membranes into graphene oxide layers arranged in close parallel alignment, the filtration capability for specific substances such as pollutants is significantly enhanced [85, 86]. This enhancement arises from the nanometer-scale interlayer spacing in graphene membranes, which is extremely narrow and arranged with an offset configuration, enabling the rejection of very small molecules [87]. However, this configuration can impede water flow, reducing permeability efficiency.

Therefore, tuning the interlayer spacing of graphene oxide membranes has become crucial to achieving membranes with high permeability and antifouling properties [88, 89]. It is known that when the interlayer spacing reaches a critical value of approximately 0.8 nanometers (8 Ångströms), the membrane attains an optimal condition that allows smooth passage of water molecules while effectively blocking salt ions from permeating [90].

Alongside the advancing research on carbon-based nanomaterials such as CNTs, CQDs, and GO, which demonstrate high potential as efficient modern desalination membranes, several development studies have been conducted [49]. These include the fabrication of ultrathin free-standing carbon-based desalination membranes and the integration of carbon-based composite membranes.

## Carbon-Based Free-Standing Membranes

Free-standing ultrathin membranes possess the capability to combine the advantageous properties of both macro- and nanoscale materials within a compact and efficient structure [91, 92]. Unlike conventional membranes that typically require a supporting layer to maintain mechanical integrity, self-standing membranes can operate independently without the need for additional substrates [93]. The reduction or elimination of support layers helps to remove diffusion barriers and extra resistances, which are often the primary causes of performance degradation in desalination membranes [94]. This degradation is frequently attributed to the phenomenon of Internal Concentration Polarization (ICP), where salt accumulation occurs on the feed side of the membrane, ultimately reducing the clean water flux and separation efficiency [95].

Therefore, designing fully free-standing membranes offers an opportunity to enhance separation process efficiency by minimizing mass transport resistance and increasing water flux [96]. Additionally, the ultrathin nature of these membranes further supports increased permeability due to the significantly shortened diffusion path for water molecules, while simultaneously enabling more precise control over molecular selectivity through nano-structural engineering [97]. The combination of good mechanical properties, high permeability, and potential for high ion selectivity makes free-standing membranes highly promising for applications such as seawater desalination, industrial wastewater treatment, and large-scale purification of clean water [98, 99].

### CNT-Based Free-Standing Desalting Membrane

Recently, flexible films assembled from CNTs, known as buckypaper, have emerged as the predominant form of free-standing desalination membranes due to the relatively complex fabrication process of vertically aligned CNTs [100]. Buckypaper generally consists of a tangled network of CNT bundles and fibers, along with some individual tubes, where CNTs are held together by non-covalent interactions without permanent chemical bonding [101, 102]. The Young's modulus, thermal conductivity, and electrical conductivity of these membranes can be enhanced by controlling the arrangement and orientation of the carbon nanotubes [103, 104].

Buckypaper (BP) has been extensively studied for various applications. Further research has explored its potential as a filtration membrane [105]. Rashid et al. fabricated BP membranes from a dispersion solution comprising multiwall carbon nanotubes (MWCNTs) and biopolymers [106]. The mechanical properties of the free-standing CNT membranes improved as the biopolymer adhered to the surfaces of short nanotubes, effectively binding them together [100]. This modification made the membranes promising candidates for water filtration and solute separation.

Additionally, the desalination performance of MWCNT membranes combined with chitosan as a surfactant was analyzed by Alsuhybani et al. in four types of salt solutions (NaCl, $Na_2SO_4$, $MgSO_4$, and $MgCl_2$) [107]. The membranes exhibited high mechanical strength with a tensile strength of 28 ± 1 MPa. They demonstrated higher salt rejection efficiency for NaCl compared to $MgSO_4$ under highly acidic conditions [107].

Dumée et al. fabricated CNT BP membranes featuring a high water contact angle of 113°, porosity up to 90%, and relatively low thermal conductivity of 2.7 kW/m$^2$ h using a vacuum filtration method. The desalination performance was tested in a Direct Contact Membrane Distillation (DCMD) configuration. The membranes showed a salt rejection rate of 99% and a permeate flux of 12 kg/m$^2$ h at a water vapor pressure of 22.7 kPa. Furthermore, a PTFE layer was added to the top surface of the CNT-BP membrane to enhance hydrophobicity, thereby preventing water absorption into the membrane and potentially reducing temperature polarization on both membrane surfaces. The contact angle increased by 23%, and the bubble point pressure rose by 28%. Both durability and water vapor permeability improved, with salt rejection exceeding 99% in the DCMD system [108].

This study clearly demonstrates that enhancing hydrophobicity leads to improved vapor transport, likely related to the wetting of membrane pore surfaces and the formation of a thin air layer acting as an insulating barrier at the membrane surface due to hydrophobic effects. This phenomenon is believed to help reduce temperature polarization in the boundary layer and minimize heat diffusion within the membrane structure.

Composite membrane from nanocellulose (NC) networks and carbon nanostructures (CNS) that can stand independently. CNS consists of a CNT network formed through branching and cross-linking during chemical vapor deposition. The NC/CNS membrane containing 50% NC maintained excellent electrical conductivity, with an ion rejection efficiency of 60% for $MgSO_4$. Due to its high electrical conductivity, this free-standing membrane also holds potential to reduce fouling via electrically assisted approaches (Fig. 1).

## GO-Based Free-Standing Desalting Membrane

Porous two-dimensional (2D) graphene layers with pore sizes ranging from 7.6 nm to 1 μm, fabricated using chemical vapor deposition (CVD), have demonstrated significantly improved permeability performance for gases, liquids, and water vapor compared to widely used commercial membranes [110]. This enhancement is attributed to the presence of nanoscale pores in graphene, which play a crucial role in ion selectivity, salt rejection efficiency, and water transport rates [111].

Free-standing GO membranes composed of stacked two-dimensional GO or reduced graphene oxide (rGO) nanosheets are considered among the most promising candidates for future desalination applications due to their unique structural characteristics [112]. Stacked GO nanosheets can form membranes in which the interlayer nanochannels act as selective transport pathways, allowing only gases and small ions to pass while rejecting larger species [113].

This has been demonstrated an efficient method to fabricate ultrathin, self-supporting rGO membranes with thicknesses ranging from 20 to 200 nm using hydroiodic acid (HI) vapor treatment and water-assisted delamination. Filtration tests using forward osmosis (FO) showed that these thin yet robust rGO membranes not only mitigated internal concentration polarization (ICP), but also provided high rejection of both organic and inorganic contaminants [114].

The narrow interlayer spacing between GO sheets is critical for enabling water transport. Therefore, various strategies have been developed to introduce additional nanochannels within GO laminates to enhance water flux while maintaining high rejection efficiency. Nonetheless, practical implementation of layered GO membranes in desalination remains challenged by issues such as scalability, interlayer swelling upon water exposure, and limited mechanical stability under high pressure.

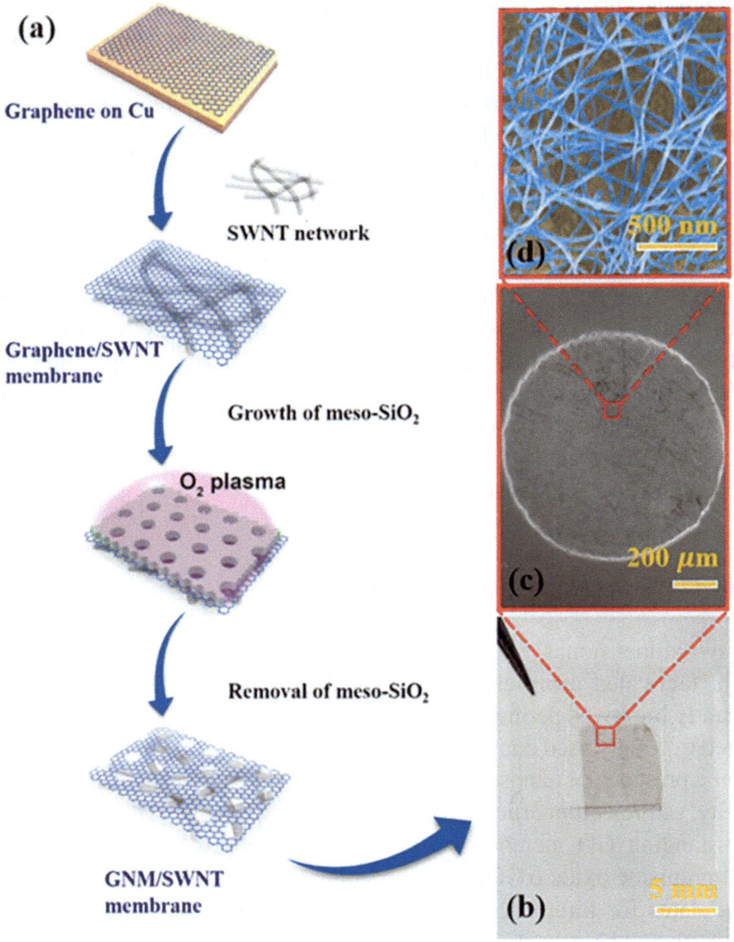


**Fig. 1** Fabrication of GNM/SWNT hybrid membranes and structural characterizations. (**a**) Schematic illustration of the process used to fabricate GNM/SWNT hybrid membranes. (**b**) Photograph of the GNM/SWNT hybrid membrane suspended on a porous PDMS substrate with 16 holes (1 mm in diameter). (**c**) SEM image of a single hole in (**b**). (**d**) Magnified SEM image of the selected red area in (**c**)

The spacing between adjacent carbon planes plays a key role in determining GO membrane separation efficiency [115].

Compared to composite membranes, free-standing membranes based on carbon nanomaterials generally exhibit superior desalination performance. This is primarily due to the reduced impact of ICP resulting from the absence of support layers, thereby enhancing separation efficiency. Numerous studies have demonstrated the effectiveness and strong potential of carbon-based free-standing membranes for seawater desalination. However, large-scale production and membrane swelling in aqueous environments remain critical challenges that must be addressed for industrial application [116].

## Carbon Incorporated Composite Membranes

Composite membranes are engineered to enhance the performance of conventional membranes, particularly polymer membranes commonly used in water treatment. Since the 1980s, polymer-based membranes have been the preferred choice for separation applications, especially seawater desalination, due to their well-developed technology, straightforward fabrication processes, low production costs, and reasonably good permeability and salt rejection performance [117].

Generally, carbon nanomaterial-based composite membranes utilize a polymer matrix as the base material and are synthesized via methods such as interfacial polymerization and phase inversion [118]. A primary limitation of conventional polymer membranes is their relatively low water permeability [119]. Therefore, carbon nanomaterial-based composite membranes offer an effective solution to increase permeability without compromising salt rejection efficiency [29]. Uniform integration of carbon nanomaterials into the membrane matrix significantly enhances permeability. In this system, carbon nanomaterials function as additives that reinforce the physical and chemical properties of the primary membrane [106].

One major contribution of carbon nanomaterials in composite membranes lies in their superior mechanical properties, especially high axial strength and stiffness resulting from their anisotropic structures [120]. Consequently, the presence of carbon nanomaterials can improve mechanical durability and long-term stability of membranes, particularly under high-pressure reverse osmosis conditions (≥15 bar) [121]. Moreover, this integration helps reduce water flux decline caused by membrane compression and increases chemical resistance against aggressive agents such as chlorine, which typically degrade polymer layers over time [122].

Research on carbon nanomaterial integration within polymer membrane structures shows promising prospects. This strategy supports the development of small-scale desalination systems with reduced energy consumption [29, 123]. One important innovation in membrane technology is the development of thin-film composite (TFC) membranes, first introduced by Cadotte in the 1970s via interfacial polymerization. TFC membranes typically consist of three layers: a non-woven fabric substrate for mechanical support, a thin porous middle layer, and a dense, highly selective top layer [124].

Carbon nanomaterials such as graphene, CNT, and GO can be incorporated into various parts of TFC membranes to improve pore structure, mechanical strength, and separation selectivity [125]. Additionally, the development of mixed matrix membranes (MMM) or hybrid membranes offers an attractive alternative. These membranes combine the unique characteristics of polymer matrices and inorganic particles to create new materials with enhanced performance [126].

Common fabrication methods for polymer membranes include phase inversion, interfacial polymerization, stretching techniques, track-etching, electrospinning, and layer-by-layer assembly [127]. Besides polymer membranes, inorganic membranes also exhibit advantages due to their high chemical and physical resistance, tunability, and reusability potential. Inorganic membranes, such as ceramic membranes, are typically constructed as multilayer structures with thin selective layers deposited on porous substrates made from metal oxide powders like alumina, silica, or titania [128]. Moreover, carbon nanomaterials have also been applied as fillers in ceramic membranes to improve mechanical properties, thermal stability, and separation efficiency [129].

Over the past decade, numerous studies have explored the integration of carbon nanomaterials with existing modern desalination membranes. Based on the working mechanism or driving force used in the separation process, composite membranes are broadly classified into two types: pressure-driven membranes and thermal-driven membranes [130].

### Pressure-Driven Membranes

The separation process using pressure-driven membranes is a technique that utilizes pressure as the main driving force to separate permeate from retentate. The membranes used in this process are known as pressure-driven membranes. This category includes reverse osmosis (RO), nanofiltration (NF), and forward osmosis (FO) membranes integrated with carbon-based nanomaterials (Fig. 2) [131, 132].

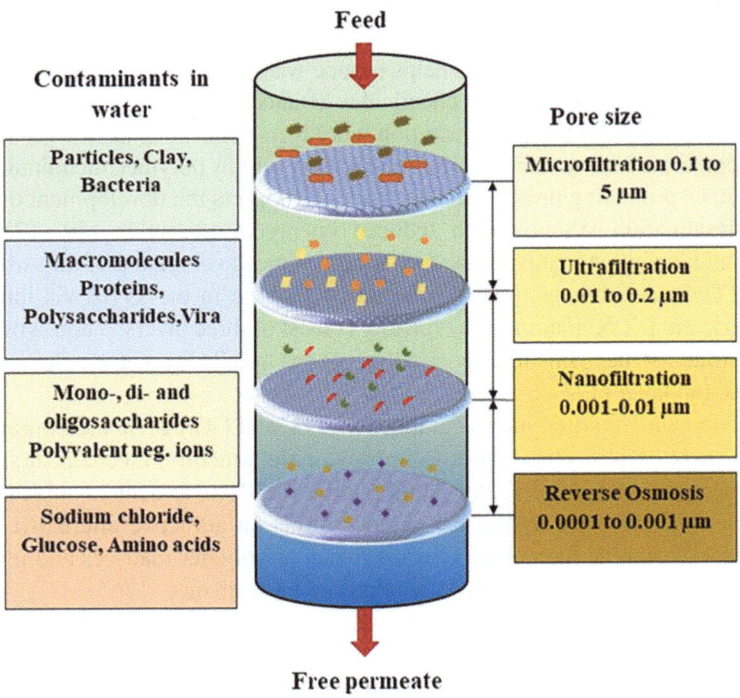


**Fig. 2** Schematic illustration of pressure-driven membrane filtration processes [133]. Reproduced from ref. with permission [133] from Elsevier, copyright 2018

### *Reverse Osmosis (RO) Membranes with Carbon-Based Nanomaterials*

Reverse osmosis (RO) is a desalination technology that relies on high pressure to separate water from saline solutions. RO offers several advantages, including low chemical usage, modular design, ease of operation, and compatibility with other technologies. However, its main challenges include membrane fouling, high energy consumption, and the trade-off between salt rejection and water flux [134]. The incorporation of CNTs with controlled pore size and surface chemistry has been proven to improve both permeability and solute rejection. Sabir et al. fabricated membranes from a citric acid and PEG400 mixture modified with MWCNT-COOH, showing enhanced salt rejection and flux at 4.0 bar pressure [135].

Most carbon-based RO membranes are fabricated using phase inversion, with small additions of carbon nanomaterials into the polymer matrix to enhance performance without sacrificing selectivity. Selected recent studies on RO membranes integrated with carbon-based nanomaterials are summarized in Table 3 [136].

### *Nanofiltration (NF) Membranes with Carbon-Based Nanomaterials*

Nanofiltration (NF) membranes are water filtration technologies characterized by their superior ability to reject divalent ions such as calcium and magnesium, while maintaining higher water flux compared to reverse osmosis (RO) membranes [142]. A key innovation in NF membrane development is the integration of carbon-based nanomaterials like CNTs and graphene, which possess unique surface properties and extremely small, uniform pore structures. The addition of these carbon nanomaterials has been proven to significantly enhance filtration capacity [29, 30]. For example, surface modification of CNTs with hydrophilic groups improves CNT dispersion within the polymer matrix, resulting in membranes with better water permeability and superior filtration performance [61]. Positively charged NF membranes fabricated by combining amine-based compounds and CNTs also demonstrate high multivalent ion retention, alongside enhanced hydrophilicity, mechanical stability, and fouling resistance [143].

A major challenge in NF membrane application is flux decline caused by fouling, or the accumulation of contaminants on the membrane surface [144]. Approaches to mitigate this issue include enhancing membrane surface hydrophilicity to form an effective water molecular layer that inhibits contaminant deposition [145]. For instance, blending functionalized MWCNTs with materials such as $TiO_2$ and ZnO into the polymer matrix improves antifouling properties and reduces membrane surface roughness [146]. Additionally, superhydrophobic membranes coated with MWCNTs and polydimethylsiloxane (PDMS) effectively decrease membrane-water surface interactions, thereby minimizing fouling [147].

Besides CNTs, graphene and GO are prominent materials in NF membrane development due to their hydrophilic nature and negatively charged nanochannels that enable rapid water flow while rejecting salt ions and organic contaminants [125]. GO modification with carboxyl groups enhances membrane performance by improving salt rejection and permeate flux. Carboxylated GO membranes combined

**Table 3** Recent studies on RO membranes integrated with carbon-based nanomaterials

| Membrane matrix | Carbon nanomaterial(s) | Method (how integrated) | Membrane performance (flux/permeance) | Salt rejection (reported) | References |
|---|---|---|---|---|---|
| Porous polymer support + nanostructured carbon (a-C:N) active film (NC-based membrane) | Nanostructured amorphous carbon with nitrogen doping (HiPIMS-deposited carbon film) | HiPIMS (high-power impulse magnetron sputtering) to deposit ultrathin NC films directly on polymer substrates (PVP sacrificial layer technique described) | High water permeance—authors report ~21.6–22.9 L $m^{-2}$ $h^{-1}$ $MPa^{-1}$ (0.2 wt% NaCl tests) and very high flux at low pressures compared to previous carbon films | ≈96% rejection for 0.2 wt% NaCl (best composition); performance varies with Na content and thickness | [137] |
| Cellulose acetate (CA) asymmetric RO membrane (CA active/dense layer) | Graphene oxide (GO) nanosheets (low wt% additions) | Phase-inversion/casting of CA casting solution containing GO; post-treatment and bench-scale RO testing at 25 bar | Generally small change or slight flux improvement depending on loading; authors report permeate flux comparable to or slightly higher than pristine CA (depends on sample) | At 1.0 wt% GO authors reported ~1.8× improvement in NaCl rejection compared to pristine CA at 25 bar (exact SR and flux values in paper) | [138] |
| Cellulose acetate/PEG400 matrix with surface-engineered MWCNT (SE-MWCNT)tethered | Surface-engineered MWCNTs (SE-MWCNT, carboxylated/tethered) | Tethering/dispersion of SE-MWCNT into CA/PEG400 matrix; bench-scale RO pilot testing (1000 ppm NaCl feed in pilot unit) | Permeate flux reported for tested configurations (see paper for flux vs. loading); membranes showed stable permeation in pilot tests | 99.8% salt rejection reported for the highest SE-MWCNT content (1000 ppm NaCl feed) | [135] |

| | | | | | |
|---|---|---|---|---|---|
| Crosslinked PVA (PVA/DGEBA)thin film (stand-alone RO) | MWCNTs (MWCNTs functionalized) + Pluronic F-127 surfactant; variants also with GO in later works | Mixed-matrix thin-film fabrication: PVA crosslinked with DGEBA; MWCNTs + Pluronic F-127 incorporated into active layer (dead-end and cross-flow RO tests reported) | Authors report substantial increases in water permeability and improved chlorine/biofouling resistance vs neat PVA; flux numbers depend on composition and test cell | High salt rejection maintained in optimized samples (authors report excellent RO performance; see Desalination 2016 article for numeric SR) | [139] |
| PVDF/PVDF-based support with GO doping + polydopamine (PDA) top layer (dual-layer) | Graphene oxide (GO) (doped) + polydopamine (PDA) coating | GO doping into PVDF support/active layer followed by PDA deposition (coating) to make dual-layer membrane | Very high pure water flux reported for the prepared membrane in filtration tests—$J_w$ = 188 LMH/barreported in the paper (note: check testing configuration in article) | Effective heavy-metal removal demonstrated: e.g., Pb(II) treatment capacity reported as 2106.36 L $m^{-2}$ (treatment volume before breakthrough) in the cited study | [140] |
| Nanoporous carbon composite membranes (porous carbon fiber layer on porous substrate)—multi-process (including RO tests) | Nanoporous carbon/porous carbon fibers (all-carbon or carbon-composite constructs) | Growth/deposition of porous carbon (carbon fiber) layer on porous ceramic/polymer substrate; tested in vacuum MD, FO and RO lab setups | Authors report very high freshwater flux (3.5–20× higher than many polymeric membranes in lab tests) in multiple processes; absolute flux depends on configuration | Reported ~100% (or near-complete) salt rejection in lab RO tests for some configurations (mechanism: interfacial salt sieving/surface transport) | [141] |

with polymers such as PPSU have been successfully used at the laboratory scale for efficient removal of heavy metal ions, achieving over 98% removal of anions and approximately 80% for cations, while maintaining high water flux. Introducing amine groups on GO nanoparticles also improves membrane surface roughness, significantly increasing pure water flux.

However, a critical challenge remains in producing ultrathin, high-quality graphene-based membranes at a large scale, which limits commercial application. Nevertheless, ongoing research highlights the great potential of carbon nanomaterials to enhance NF membrane performance for efficient water filtration and contaminant removal. Modifying NF membranes with GO and its derivatives has demonstrated significant performance improvements. Incorporating hybrid GO into membranes increases surface hydrophilicity, contributing to enhanced removal of contaminants such as copper ions and dyes. Combining GO with materials like chitosan results in membranes with higher hydrophilicity, lower surface roughness, and improved water flux and salt rejection compared to current commercial membranes. The nanochannel structure between graphene sheets also plays a vital role in accelerating water permeation and reducing mass transfer resistance.

To enhance the stability and size of these nanochannels, various spacers such as silver nanoparticles, $MoS_2$, and magnesium silicate are inserted or cross-linked with GO sheets, reinforcing membrane structure and preventing layer collapse. Incorporating nanodiamonds into GO layers yields membranes with larger interlayer spacing and increased porosity, improving filtration rate and selectivity toward multivalent ions. However, the extremely narrow interlayer spacing of GO (~0.36 nm) can hinder water flow, which is addressed by inserting nanomaterials like CNTs. Oxygen-functionalized CNTs strengthen electrostatic interactions between the membrane and charged ions, boosting NF membrane ion rejection. Composite membranes combining rGO and CNT exhibit significant improvements in permeability and selectivity.

Additionally, various fabrication techniques including vacuum filtration, layer-by-layer (LbL) assembly, and inkjet printing have been employed to produce thin, uniform GO membranes with excellent water flux and organic molecule rejection capabilities. Zero-dimensional carbon nanomaterials such as GOQDs also show great promise in enhancing membrane hydrophilicity, mechanical properties, and antibacterial performance. GOQDs incorporated in nanocomposite membrane layers create efficient water permeation pathways, increasing water flux without compromising solute retention.

The use of carbon nanomaterials also improves membrane fouling and chlorine resistance, as well as mechanical stability. Beyond graphene and CNTs, metal–organic framework (MOF)-based composite membranes embedded in GO layers demonstrate enhanced water flux and effective heavy metal ion rejection.

Overall, advances in carbon nanomaterial-based membrane technology offer significant opportunities for more efficient and selective water filtration applications. Selected recent studies on NF membranes integrated with carbon-based nanomaterials are summarized in Table 4.

**Table 4** Recent studies on NF membranes integrated with carbon-based nanomaterials

| Membrane matrix | Carbon nanomaterials | Method | Flux/permeance | Salt rejection/ performance | References |
|---|---|---|---|---|---|
| Polyamide (PA) thin-film composite (TFC) | Graphene oxide (GO), 0.2 wt% | GO embedded via interfacial polymerization | Water flux increased up to 12× compared to pristine PA | Similar NaCl rejection (~95%) maintained | [148] |
| Piperazine (PIP)-based PA NF membrane | Graphene oxide (GO) | In-situ interfacial polymerization (PIP–GO + TMC) | +10–15 LMH higher water flux | Enhanced divalent ion rejection ($Ca^{2+}$, $Mg^{2+}$) | [149] |
| Polyethersulfone (PES)/ PA composite (TFC) | Double-walled CNTs (DWCNT) | Interlayer deposition + interfacial polymerization | ≈20–21 $L\ m^{-2}\ h^{-1}\ bar^{-1}$ | Salt rejection >98% for divalent ions | [150] |
| Graphene oxide (GO) nanosheet membranes | Holey GO/reduced holey GO (r-hGO) | Controlled pore formation on GO (etching + reduction) | Water permeance significantly increased (several-fold vs. pristine GO) | Maintained NF-level ion rejection | [151] |
| Polyamide TFC NF membrane | Functionalized MWCNT (–$NH_2$/–COOH) | Embedding functionalized CNTs via interfacial polymerization | Flux increased by ~25–40% depending on loading | Improved ion and dye rejection; enhanced antifouling | [152] |
| GO–Cellulose nanocomposite | GO + Cellulose nanocrystals (CNC) | Layered casting/vacuum filtration | Flux up to 30–40 LMH/bar, higher than pure GO or cellulose | >90% rejection for dyes and divalent ions | [153] |
| Commercial NF membrane (NF270-Ppy-CNT) | Carbon nanotubes (CNTs) grown in situ | Plasma-assisted CNT growth on membrane surface | Improved flux and NaCl permeability tuning | NaCl rejection significantly increased | [154] |

*Forward Osmosis (FO) Membranes with Carbon-Based Nanomaterials*

Forward osmosis (FO) membranes utilizing carbon-based nanomaterials, such as CNTs, exhibit high salt rejection capabilities. The cylindrical nanostructure of CNTs aids in effectively retaining salt ions, forming the basis for high-performance FO membranes. Many studies focus on incorporating CNTs into the active layer to fabricate thin film nanocomposite (TFN) membranes. For example, Song and colleagues developed a dual-layer FO membrane by integrating dopamine and CNTs during polymerization, resulting in a membrane with increased porosity and a denser selective layer. This design reduces ICP effects and enhances osmotic water flux by up to 54% compared to traditional membranes.

Additionally, Zou et al. functionalized CNTs with zwitterionic molecules and incorporated them into the active layer of commercial FO membranes. This modification significantly decreased reverse solute flux, especially when CNTs were embedded in the active layer rather than the support layer. Meanwhile, Cui and his team blended MWCNTs into cellulose acetate membranes, producing membranes with increased hydrophilicity and approximately 50% higher water permeability. These membranes also exhibited improved fouling resistance against substances like alginate due to enhanced electrostatic repulsion.

Further advancements include the fabrication of ultrathin carbon nanomembranes (CNMs) about 1.2 nm thick, which allow extremely rapid water passage while effectively blocking ions and even protons. These membranes transport water twice as fast as commercial FO membranes and rapidly reduce salt concentrations in the draw solution.

Other studies have employed functionalized carbon nanofibers with chemical groups such as carboxyl and amine, increasing membrane hydrophilicity and more than doubling osmotic water flux compared to pure cellulose membranes, while maintaining very low reverse salt flux. A common challenge in large-scale graphene membrane production is the presence of defects. However, phase inversion techniques combined with plasma treatment have minimized defects, enabling control over pore size and membrane selectivity. These graphene membranes can selectively separate salt ions with high precision.

In GO-based membranes, controlling interlayer spacing is crucial. Jin and colleagues fabricated membranes with ultra-thin, dense GO layers, achieving high water flux, salt rejection up to 99.9%, and resistance to harsh chemical environments. GOQDs aid in linking GO sheets, strengthening and densifying membrane structure. Nitrogen doping in graphene membranes reduces reverse salt leakage, while adding specific surfactants between GO layers increases interlayer spacing and water flow. Cross-linking agents such as ethylenediamine and polyacrylic acid further enhance spacing and membrane performance. Chemical reduction of GO to remove hydrophilic groups narrows nanochannels, improving membrane selectivity. Adding a hydrophilic polydopamine (PDA) layer on reduced GO surfaces also improves water permeability.

CNT-based membranes have also been used as alternative selective layers with porous, easily wetted hollow fiber substrates, reducing ICP and significantly increasing water flux and salt rejection. Furthermore, developing 3D porous polymeric GO support layers with interconnected structures provides an effective solution to reduce ICP, boosting water flux by up to 73% compared to conventional one-dimensional (1D) support layers.

In conclusion, the use of carbon nanomaterials in FO membranes is rapidly evolving with various strategies to enhance water flux, salt rejection, fouling resistance, and chemical stability. These membranes hold significant potential for highly efficient desalination and water purification applications. Selected recent studies on FO membranes integrated with carbon-based nanomaterials are summarized in Table 5.

## Thermal-Driven Membranes

Thermal-driven membrane processes involve desalination methods based on evaporation, where thermal energy is used to heat water. The heated water evaporates, and the vapor is subsequently condensed back into clean, pure water. Due to the relatively high energy consumption, these processes are more suitable for use in power plants or oil refineries, where waste heat can be effectively utilized as an energy source. Thermally-driven membranes include membrane distillation (MD) and pervaporation (PV) membranes integrated with carbon-based nanomaterials.

### *Membrane Distillation (MD) with Carbon-Based Nanomaterials*

The structure of MD membranes critically affects mass and heat transfer efficiency. Key membrane properties include high hydrophobicity, large porosity, uniform pore size, low tortuosity, and thin thickness to optimize performance. Recently, the incorporation of CNTs into membranes, known as carbon nanotube incorporated membranes (CNIM) has shown to significantly enhance desalination efficiency. CNIMs exhibit higher water flux and improved salt rejection compared to conventional membranes, allow operation at lower temperatures, and reduce salt concentration polarization effects.

Hydrophilic functionalized CNTs (e.g., CNT-COOH) improve vapor permeation without wetting pores. Dual-layer membranes with hydrophobic CNT layers on the feed side and hydrophilic CNT layers on the permeate side achieve very high fluxes, reaching up to 126 kg/m$^2$ h. CNTs embedded in PVDF membranes create porous sponge-like structures, resulting in high flux with excellent salt rejection.

Fouling remains a main challenge in MD. Hydrophobic coatings prevent membrane wetting. In-situ CNT coatings can form superhydrophobic, highly porous surfaces. Vertically grown CNT forests (CNTF) on porous substrates generate rough surfaces with open pores, enhancing flux and fouling resistance. The electrical conductivity of CNTs enables active fouling control by applying an electric field, which increases local pH, accelerates flux recovery, and prevents salt scaling. Electrospun membranes with CNT additives form superhydrophobic, highly porous structures

**Table 5** Recent studies on FO membranes integrated with carbon-based nanomaterials

| Membrane matrix | Carbon nanomaterial | Method | Flux/permeance (reported) | Salt rejection/ performance (reported) | References |
|---|---|---|---|---|---|
| Cellulose triacetate (CTA) FO membrane (TFC-type tests) | Graphene oxide (GO) nanosheets (aqueous-phase addition, low loadings ~0.01 wt%) | GO dispersed into aqueous phase during interfacial polymerization (IP); TFC FO membrane fabrication | Water flux (AL-FS, draw 2.0 M NaCl): increased from 7.32 → 11.49 L $m^{-2}$ $h^{-1}$ (≈57% increase) at optimum GO loading; AL-DS: 10.17 → 14.49 L $m^{-2}$ $h^{-1}$ | Selectivity maintained; improved surface hydrophilicity and chlorine resistance; moderate changes in RSF | [155] |
| Cellulose triacetate (CTA)/CTA blends | Graphene oxide (GO) nanosheets incorporated into casting solution/ active layer | GO incorporated into casting solution or active layer (casting/IP) | Reported significant flux improvement vs. neat CTA; specific flux depends on GO loading and drying; e.g., several studies report 20–60% flux increase | Improved salt rejection stability and reduced fouling vs. pristine CTA | [156] |
| Polymer supports (PES/PSf) or support-free laminar GO membranes | Reduced/holey graphene oxide (r-GO/hGO) laminar membranes | Laminar stacking (vacuum filtration/layer-by-layer) with in-plane nano-holes (etching) or support-free molecular layer-by-layer (mLbL) assembly | Holey/r-GO FO membranes report several-fold increases in water permeance versus pristine GO laminates (exact values depend on hole size, thickness) | Maintained or improved salt rejection/selectivity; higher water flux while keeping low reverse solute flux in optimal designs | [157] |
| FO membranes with activated carbon/ carbonaceous additives | Activated carbon (powder), carbon black, functional carbon additives | Incorporation in casting solution or coating; composite porous supports | Reported improved hydrophilicity and water flux compared to unmodified supports; example: notable flux and fouling resistance improvements in CTA blends | Improved salt rejection stability and lower fouling propensity | [158] |

| | | | | | |
|---|---|---|---|---|---|
| Nanoporous carbon composite membranes (carbon-fiber/ porous carbon films) | Nanoporous carbon (porous carbon fiber layers/carbon composite) | Porous carbon layer grown/ deposited on porous substrate (composite membrane) | Very high water flux reported (3.5–20× higher vs. polymeric membranes in some tests); specific FO fluxes depend on configuration | Reported near-complete salt rejection in lab tests (mechanism: interfacial salt sieving/surface transport) | [141] |
| Modified PES/PSf supports with graphene derivatives | Graphene oxide (GO) or reduced GO (rGO) dispersed in support/top layer | GO dispersed in support by phase inversion or casting; or thin GO coating via vacuum filtration | Examples report increased FO water flux (e.g., 10–50% improvement depending on loading) and decreased internal concentration polarization (ICP) due to more porous/hydrophilic supports | Improved water flux with comparable or better solute retention/lower RSF | [159] |

ideal for MD. Simulations also indicate graphene mesoporous membranes outperform conventional membranes in flux.

GO composites with PVDF are widely used in AGMD and DCMD configurations, offering chemical stability, wetting resistance, high flux, and near-complete salt rejection. GO modified with additives like APTS and PVP facilitates vapor transport through graphene gaps and microscopic channels, increasing contact angles and vapor diffusivity. GO functionalized with FTES or NBA improves hydrophobicity, liquid entry pressure (LEP), and mechanical strength. For example, PVDF membranes with 0.5 wt% GO-NBA exhibit increased tensile strength and pressure resistance.

In vacuum membrane distillation (VMD), graphene can act as a Joule heater, improving energy efficiency. Simulations show graphene prevents fouling by keeping water channels open and inhibiting salt accumulation. Graphene nanoplatelets (GNP) coated on pilot-scale PE membranes reduce fouling by up to 78% after 10 h. Zero-dimensional nanoparticles like carbon quantum dots (CQD) enhance antifouling and superhydrophobic properties. For instance, GQD/PVDF membranes in AGMD achieve high flux and near-perfect salt rejection in long-term tests. Activated carbon (AC), used as large particles or nanofiber structures, provides large pores to enhance vapor pathways and filtration performance. Some configurations demonstrate fluxes with over 99% salt rejection.

Overall, integrating carbon nanomaterials such as CNT, GO, GNP, CQD, and AC into common polymer membranes like PVDF, PTFE, and PP significantly improves flux, stability, and salt rejection efficiency in MD processes, even with minimal additive loading. Selected recent studies on MD membranes incorporating carbon-based nanomaterials are summarized in Table 6.

*Pervaporation (PV) Membranes with Carbon-Based Nanomaterials*

Membrane performance in pervaporation (PV) processes is generally evaluated based on flux and separation factor of membrane, where flux is defined as the mass transfer per unit membrane area over a specific time period. Additionally, the separation factor measures the effectiveness of component separation in a mixture based on the molar fractions in the feed and permeate.

Although the utilization of CNTs in PV membranes remains limited (see Fig. 3), Yang et al. successfully developed a MMM based on PVA with MWCNTs. Strong interactions between PVA and MWCNTs occurred through interfacial adhesion, hydrogen bonding, and covalent bonding. The resulting PVA/C-MWCNT/MA membrane exhibited excellent desalination performance, achieving a salt rejection of 99.91% and a water flux of 6.96 kg $m^{-2}$ $h^{-1}$ in a 35,000 ppm NaCl solution at room temperature, along with high stability over a 30-h durability test.

GO-based membranes also demonstrate significant potential for desalination applications. The nanochannels in GO layers, stabilized by cross-linking cations, allow selective transport of water molecules while hindering salt ions such as $Na^+$. Increasing temperature accelerates water permeation without affecting ion migration, a phenomenon not observed in non-thermal systems. Compared to

**Table 6** Recent studies on MD membranes integrated with carbon-based nanomaterials

| Membrane matrix | Carbon nanomaterial | Method | Application | Flux (reported) | Salt rejection (reported) | References |
|---|---|---|---|---|---|---|
| PTFE membrane immobilized with multi-walled CNTs (CNIM) | Carbon nanotubes (CNT) immobilized on membrane surface | CNT immobilization (drop-casting/adhesion) on commercial hydrophobic membrane; tested in DCMD | Treatment of high-salinity produced water/ desalination; anti-scaling & antifouling studies | Water vapor flux **increased up to ~77%** vs pristine membrane; CNIM regained ~90.9% initial flux after washing (improved flux stability) | Very high salt rejection (reported ≈ **>99.9%**; negligible permeate conductivity increase) | [160] |
| PVDF mixed-matrix membrane with reduced graphene oxide (r-GO) | Reduced graphene oxide (r-GO) nanosheets | r-GO incorporated as filler into PVDF casting solution (mixed-matrix membrane, phase inversion) | DCMD/vacuum MD for seawater/brackish desalination | Flux ~31.9–60 $L\ m^{-2}\ h^{-1}$ reported for optimized r-GO loadings (value depends on study/processing). Example: one work reported ~31.9 $L\ m^{-2}\ h^{-1}$ (≈32 $kg\ m^{-2}\ h^{-1}$) | High salt rejection ≈99.7–99.9% (stable against wetting in optimized membranes) | [161] |
| PVDF electrospun/ POSS-functionalized GO-PVDF composite | Graphene oxide (GO)/POSS-GO | Electrospinning + nanoparticle incorporation/surface treatment | DCMD with improved wetting resistance/long-term stability | Stable flux **~28** $\mathbf{L\ m^{-2}\ h^{-1}}$ over 5 days continuous testing in reported study | High salt rejection (reported **>99%**; stable over test duration) | [162] |
| Nanoporous carbon composite membranes (porous carbon fiber layer on substrate) | Nanoporous porous carbon/ carbon fiber layer | Growth/deposition of porous carbon layer on ceramic/porous substrate (composite membrane) | Vacuum MD/general desalination—high-flux concept | Extremely high freshwater flux: **3.5–20×** higher than many polymeric membranes in lab tests; some tests report very large flux multiples (absolute flux depends on test T) | Reported **100% (or near-complete) salt rejection** in lab experiments | [141] |

(continued)

**Table 6** (continued)

| Membrane matrix | Carbon nanomaterial | Method | Application | Flux (reported) | Salt rejection (reported) | References |
|---|---|---|---|---|---|---|
| PVDF/carbon-nanotube blended flat sheet (CNT functionalized) | Functionalized MWCNT (e.g., CNT–ODA or COOH) blended into PVDF | Blending (casting) + phase inversion; sometimes plus surface coating | DCMD for saline water/produced water | Example reported maximum flux **~23 L $m^{-2}$ $h^{-1}$** for CNT-modified membrane (depending on loading) | Reported salt rejection up to **≈98–99.9%** (depending on test and loading) | [163] |
| CNT-immobilized PTFE/CNIM — scaling & fouling mitigation studies | CNT (immobilized) | Immobilization on hydrophobic membranes; DCMD tests with scaling salts ($CaSO_4$, $CaCO_3$, $BaSO_4$) | Fouling/scale reduction in MD for produced water/high scaling feeds | CNIM showed **up to 127% higher flux** than pristine membrane after 10 h in $CaSO_4$ test; normalized flux decline much lower for CNIM. | Salt rejection remained very high (close to **100%**) while scaling decreased | [164] |
| PVDF/graphene-nanoplatelet or carbon black-nanoparticle enhanced flat sheets | Graphene nanoplatelets (GNP), carbon black, carbon quantum dots (GQDs) | Filler addition into casting solution; sometimes surface spray/coating | DCMD/AGMD—enhanced hydrophobicity, anti-wetting, antifouling | Reported stable fluxes **~16–40 L $m^{-2}$ $h^{-1}$** in different studies; e.g., PVDF-G showed ~16 L $m^{-2}$ $h^{-1}$ in one report; other optimized composites gave up to ~39.8 kg $m^{-2}$ $h^{-1}$ | High salt rejection **>99%** in optimized membranes; improved wetting resistance | [165] |

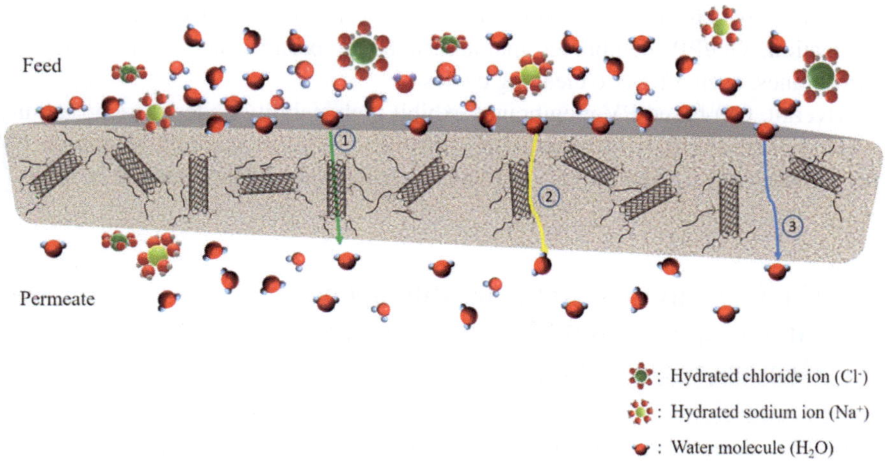


**Fig. 3** Schematic representation of water molecule transport through the PVA/C-MWCNT/MA membrane in NaCl solution: ① Transport of water through the central pores of CNT; ② Transport of water along the outer surface of CNT; ③ Water passing through the PVA matrix [166]. Reproduced from ref. with permission [166] from Elsevier, copyright 2019

pressure-driven systems, GO membranes in PV exhibit 1–2 times higher pure water flux due to the large capillary pressure.

Feng et al. developed GO composite membranes through various fabrication methods including vacuum filtration, phase inversion, and coaxial spinning, resulting in superior desalination performance. For instance, GO/polyimide mixed membranes achieved high flux (36.1 kg $m^{-2}$ $h^{-1}$) and salt rejection up to 99.9% at 90 °C. PV separation performance was further enhanced by tuning the interlayer spacing of GO. The use of cross-linking agents such as SSA, thiourea, and dendrimer PAMAM effectively controlled membrane swelling and reduced nanochannel size to approximately 0.43–0.76 nm. Consequently, permeate flux increased significantly while maintaining high salt rejection (>99.9%). Additionally, agents such as cyclohexanediamine (CDA) and *p*-phenylenediamine (pPDA) were used to create three-dimensional GO structures with larger interlayer spacing and improved surface hydrophilicity. CDA-based membranes exhibited flux of 20.1 kg $m^{-2}$ $h^{-1}$ and ion rejection of 99.9%, greatly surpassing conventional GO membranes.

Sun et al. also incorporated PVA chains into GO laminates, serving as spacers and cross-linkers. The resulting membrane (c-GO-PVA10) achieved a flux of 98.1 kg $m^{-2}$ $h^{-1}$ and salt rejection of 99.99% for 10 wt% NaCl solution at 85 °C. Another approach by Cheng et al. involved fabricating thin-film nanocomposite (TFN) membranes with robust GO layers adhered onto polyacrylonitrile (PAN) supports using PVA and glutaraldehyde. These membranes demonstrated stable flux of 69.1 L $m^{-2}$ $h^{-1}$ and salt rejection of 99.9% over 24 h. A major challenge in thermal desalination is salt scaling. Cha-Umpong et al. showed that coating GO on polypropylene (PP) hollow fiber membranes enhanced resistance to salt crystallization due to GO's smooth, hydrophilic, and negatively charged surface properties, which create electrostatic repulsion and energy barriers against salt nucleation. GO/PP membranes maintained better performance than pristine PP membranes, even after two cleaning cycles.

Overall, GO-based PV membranes exhibit high stability under extreme salinity conditions and long-term operational capability. Selected research results on PV membranes integrated with carbon-based nanomaterials are summarized in Table 7.

# 4 The Utilization of Artificial Intelligence and Computational Modelling in Carbon-Based Membrane Designing

In order to design a carbon-based membrane that corresponds with expected functions, there are several stages to be followed, as you can see in Fig. 6. The first stage is choosing base materials in dependence to the expected functions and characteristics, e.g. polymers and MOFs. The following stage is fabrication to determine its concentration, temperature, pH, humidity, and other factors, both intrinsic and extrinsic. The previous stage would produce membrane design with several characteristics, such as thickness, roughness, water contact, and angle. This product needs to be tested in different conditions, such as pressure, temperature, composition, and feeded. Scientists would then analyze the functionality (solubility, swelling, permeability, and selectivity) and theory discovery (paradigm, mechanism, and predictive model). The last stage is operational to see if there is any fouling, degradation, and plasticization. The use of artificial intelligence (AI) can be applied to material choosing, fabrication, functionality, theory discovery, and operation [173].

## *4.1 Membrane Designing Using AI Algorithms*

By using Mitchell's definition, there are three methods of machine learning algorithms for membrane science designing.

### Experience (E) in AI/ML Algorithms

Generally, there are two types of algorithms in machine learning, supervised and unsupervised, and two additional categories which have promising potential for the future of membrane sciences, weakly supervised and reinforcement learning. Unsupervised learning only providing features $x$ which have a probability output $p(x)$, while on the other hand, supervised learning providing both features $x$ and corresponding labels $y$ which have probability output $p(y|x)$. Active learning is the most common and mature method of weakly supervised learning while

**Table 7** Recent studies on PV membranes integrated with carbon-based nanomaterials

| Membrane matrix | Carbon nanomaterial | Method | Application | Flux (reported) | Separation/performance (reported) | References |
|---|---|---|---|---|---|---|
| PVA (polyvinyl alcohol) composite | GO–CNT hybrid (GO + carboxylated CNT) | Mixed-matrix composite: GO + CNT dispersed in PVA casting solution; membrane casting + crosslinking | Ethanol dehydration (pervaporation) | 0.87 kg $m^{-2}$ $h^{-1}$ (at reported conditions) | Separation factor = 523 (PVA–GO–CNT hybrid) | [167] |
| PDMS-b-PPO (block copolymer) membranes | Graphene oxide (GO) nanoparticles | GO incorporation into PDMS-*b*-PPO matrix (solution blending/casting) | Ethanol dehydration (pervaporation) | Flux range reported ~140–210 g $m^{-2}$ $h^{-1}$ (i.e., 0.14–0.21 kg $m^{-2}$ $h^{-1}$) depending on feed conc. & temp | Separation factor reported in the range ~10–150 depending on formulation and feed (check paper for condition details) | [168] |
| PVA mixed-matrix membranes (MMM) | Reduced graphene oxide (rGO) or graphene (GR)/GO | rGO/GO dispersed in PVA matrix; casting + crosslinking | Ethanol/water dehydration | GO-PVA MMMs: higher permeation flux than GR-PVA and neat PVA in many reports (typical values 0.5–1.2 kg $m^{-2}$ $h^{-1}$ depending on test) | GO-PVA: higher flux but somewhat lower separation factor than GR-PVA; GR-PVA exhibits higher separation factor but lower flux. (See paper for numbers per feed conc.) | [169] |
| Chitosan/E-VOH (ethylene-vinyl alcohol) blends | Multi-walled carbon nanotubes (MWCNT) (functionalized) | MWCNTs dispersed in polymer matrix (solution casting) | Separation of benzene/cyclohexane (aromatic/aliphatic mixtures) | Example: 1.265 kg $m^{-2}$ $h^{-1}$ total flux for a 50 wt% benzene feed in one study | Separation factor ~1.59 (depends on feed composition and temperature) | [170] |

(continued)

**Table 7** (continued)

| Membrane matrix | Carbon nanomaterial | Method | Application | Flux (reported) | Separation/performance (reported) | References |
|---|---|---|---|---|---|---|
| PVA-based mixed-matrix membranes (PVA + carbon fillers) | Graphene oxide (GO), carbon quantum dots (CQDs), carboxyl-CNT | Casting of MMMs; crosslinking (various chemical crosslinkers) | Ethanol dehydration & water removal from organic solvents | Reported flux values vary widely (typical 0.2–1.5 kg $m^{-2}$ $h^{-1}$ depending on membrane and conditions)—some GO-PVA show ~0.87 kg $m^{-2}$ $h^{-1}$ (see Gupta et al.) | Separation factors commonly in the $10^2$–$10^3$ range for optimized PVA-based MMMs (very formulation and condition dependent) | [167] |
| Functionalized polymer blends (e.g., VOH/E-VOH) with CNT | Functionalized MWCNTs | Blending and solution casting; some studies used surface functionalization to improve dispersion | Separation of aromatic/aliphatic mixtures (e.g., benzene/cyclohexane) | Flux ~0.5–1.3 kg $m^{-2}$ $h^{-1}$ in reported experiments (depends on feed and membrane) | Moderate selectivity (separation factor 1.3–3.0 typical for close-boiling organics); CNT addition improved flux and sometimes selectivity | [171] |
| Novel 2D filler/graphdiyne/MOF@GO hybrid PV membranes | Graphdiyne, MOF-303@GO, other carbon-based 2D fillers | Mixed-matrix approaches, post-synthesis incorporation of MOF@GO into polymer matrix | Ethanol dehydration/alcohol recovery | Reported permeate flux examples: 2.35 kg $m^{-2}$ $h^{-1}$ (graphdiyne MMM, recent 2024–2025 works) under specific conditions | Separation factor reported ~11.3 (for graphdiyne MMM example)—some hybrid formulations reported extremely high separation factors (>1000) for ultra-thin PEC membranes in other studies | [172] |

reinforcement learning works based on reward feedback. Reinforcement learning is still in development and faces many challenges (see Fig. 4) [173].

## The Task (T) for Membrane Design

This algorithm produces output $y$ (response) based on given data $x$ (predictor). It helps researchers to understand the relation between responses and predictors. Using that knowledge, they can use it to enhance prediction accuracy [174]. Supervised learning algorithms can be divided into two methods, classification and regression. Both of them can be analyzed further with modelling methods [174].

Classification algorithm works as a tool to classify or input the given groups of dataset into particular categories based on expected characteristics or properties. This algorithm itself is divided into several types, binary classification (yes or no categories), multi-class classification (one category over two choices of categories), and multi-label classification (multi-choices). The example of utilization of this algorithm in membrane science is the distinction of oxygen distribution in porous transport layers.

Regression is known for its purpose to determine and define the relationships between $x$ feature and output $y$, usually in the form of a mathematical polynomial equation. In the application for membrane sciences, it is typically used to predict membrane properties, the most common is modelling electrolysis polarization curves in PEM [173]. Models of supervised learning can be divided into four methods, decision trees, Gaussian process, support vector machine (SVM), and neural networks [173].

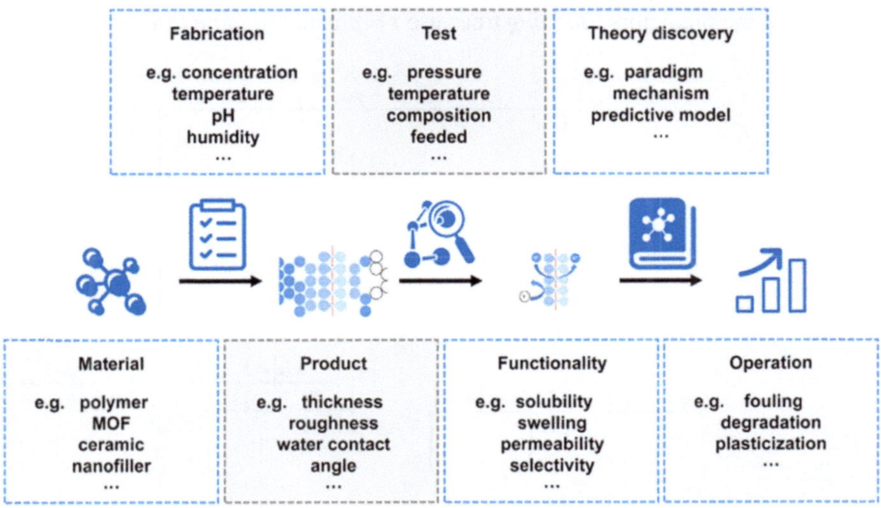


**Fig. 4** The process of membrane designing. Currently, artificial intelligence can only cover the blue parts [173]. Reproduced from ref. with permission [173] from Elsevier, copyright 2024

## Decision Tree

Decision tree methods are generally used for regression and classification. The have several advantages, such as easily explained, close to humanely behaviour in decision-making, can be displayed into graphics, and can handle qualitative predictors without the need of dummy variables, yet they also have disadvantage, which is low level of predictive accuracy. This method works by dividing or separating a given dataset into simplified spaces or regions and keeps dividing until it reaches the most specific regions, similar to branches of a tree, which is why it is known as the *decision tree* method. There are three techniques to enhance the usefullness of decision tree methods, especially for prediction. Those are bagging (bootstrap aggregation), random forests, and boosting [174]. Figure 5 shows the illustration of decision tree schematics.

*Bootstrap aggregation*, or commonly known as *bagging*, is a technique in statistical learning with a purpose to reduce the variance, which is useful and mostly used in a decision tree or tree-based methods. Fundamentally, they take a lot of training sets from the given population, create a prediction model, and make an average of the results [174]. In summary, *bootstrap aggregation* can reduce the variance by making a variation or diversity in the decision tree, where each subset is modified in such a way that it is not exactly the same as the original dataset.

*Random forests* technique is a modified version of *bootstrap aggregation* technique. This technique is supposed to consider a subset of predictors at each split. For instance, assume a split in a decision tree. At each split, there is a subset of random predictors ($r$) taken out of the full set of predictors ($p$), where $r \oplus \sqrt{p}$ is commonly used to reduce error. This way, *random forests* technique can overcome the disadvantages of *bagging* technique, where *bagging* trees are most likely to use the strong predictor in the top split while *random forests* technique is forced to consider only a subset of the predictors. *Bagging* trees use $r = p$ while *random forests* use $r = \sqrt{p}$

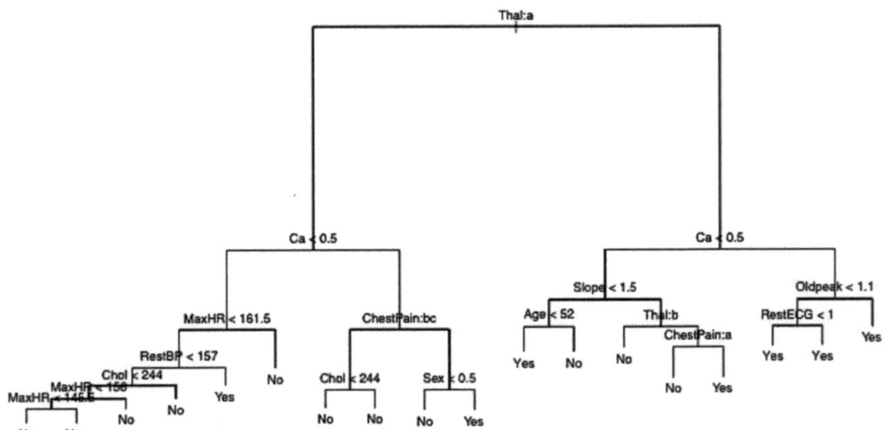


**Fig. 5** Decision tree schematics illustration [174]. Reproduced from ref. with permission [174] from Springer Science & Business Media, copyright 2013

. And on the other side, *boosting* technique works in a sequential order, where a tree is grown based on information from previous grown trees. This technique modified the original data set without involving any bootstrap sampling [174].

Gaussian Process

This model has three main stages, those are gaussian process (GP), posterior gaussian process, and GP training. This model is a stochastic process, a collection of random variables, it can be any finite number, of which have joint Gaussian distributions. On the other side, posterior GP use gaussian processes as a prior for Bayesian inference, which focus on certain properties of the function. Its main goal is to derive the simple rules of how to update this prior in the light of the training data. Finally, we can train a GP by choosing different mean and covariance functions in the light of the data. This stage would also let the hyperparameters adapt with the functions [173].

Support Vector Machine (SVM)

This model is a result of enlargement of the feature space in a specific way to form an extension of the support vector classifier, using kernels [174]. The kernels are some kind of decision machine that avoid computing posterior probabilities [176]. They describe SVM as a sparse technique, a kernel technique, and a maximum margin separator.

For this chapter, the AI and ML algorithms, most research uses Random Forests algorithm, which is categorized as supervised learning. Hence, we will not discuss a lot about the other algorithm categories.

## 5 Challenges in the Commercialization of Carbon-Based Membranes

There are not many references about the use of AI and ML in a specific carbon-based membrane designing, mostly about polymers, metal organic frameworks, mixed matrix membranes, and general membrane technology. In this chapter, we brought up some innovations in organic-based membrane designing and prediction with the help of AI and ML.

Barnett et al. use Gaussian process regression to predict two novel-high performance polymers. The polymers are translated into binary representation using a fingerprinting algorithm. One of the algorithm's advantages is it can translate a wide range of parameters, even the chemical connections. They use the translated polymers to train a ML model and test it to predict polymers over the National Institute for Material Science (NIMS) database.

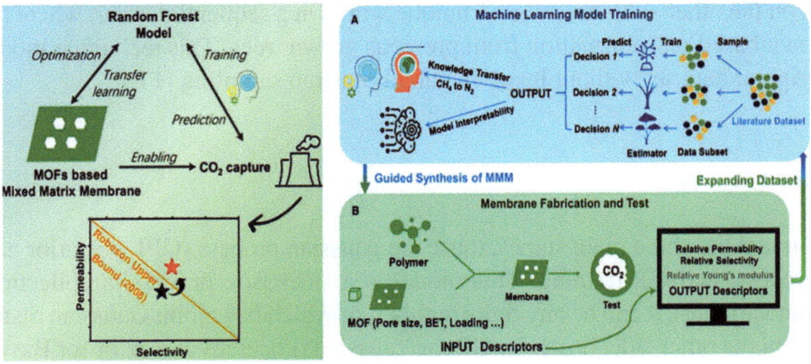


**Fig. 6** Schematic of Random Forest Model in designing MOFs-based Mixed Matrix Membrane, including the cycle of model training, fabrication, and testing [175]. Reproduced from ref. with permission [175] from Elsevier, copyright 2022

Guan et al. tried to enhance Metal Organic Frameworks (MOFs) membrane performance in carbon-capturing with the use of random forests algorithm. As you can see in Fig. 6. They use Random Forest Model to help the designing of membrane-based carbon capture. First, they train the model for RF using 648 literature datasets on Mixed Matrix Membranes (MMMs) for $CO_2/CH_4$ separation, involving 36 types of MOFs and 41 different polymers. They make four categories of variables that affecting the membrane performance, those are intrinsic properties of MOFs (types, particle size, BET surface area, aperture size, and pore diameter), properties of the polymer matrix (polymer types and control permeability and selectivity), characteristics of the MMMs (loading and thickness), and testing condition (temperature and pressure). MOF type and polymer type are considered as categorical inputs, and the rest are the numerical inputs. Their work's main goal is to quantify how relative $CO_2$ permeability and $CO_2/CH_4$ selectivity of MOFs can affect membrane performances. In order to get better predictive performance, they use cross-validation to tune the hyperparameters for robust optimization and validation. After the training of ML, they optimize MOFs using ML of literature data for $CO_2/CH_4$ separation. Some properties of polymers, such as polymer type and intrinsic gas transport rates, are important in MMMs gas separation performance, since they impact MOFs contributions to membrane performances. Using Maxwell model as reference, there is a possibility of relative permeability enhancement by blending MOF with less permeable polymer. The quality of interface between the polymer and MOF contributes significantly to gas permeation, that is why polymer types are important. Meanwhile, MOF types turn out not as important as polymer types. You can see how the ML algorithm categorized and made graphs to show some properties of the literature data in Fig. 7.

Based on the results of RF models, they chose two representative MOFs, named Cu-THQ and Cu-CAT-1, since both of them have suitable pore size with different surface area, and haven't been used in gas separation. Additionally, they have excellent stability and environmentally benign synthesis since they use water as solvent rather than organic solvent, which mostly MOFs use. They found out that Cu-CAT-1

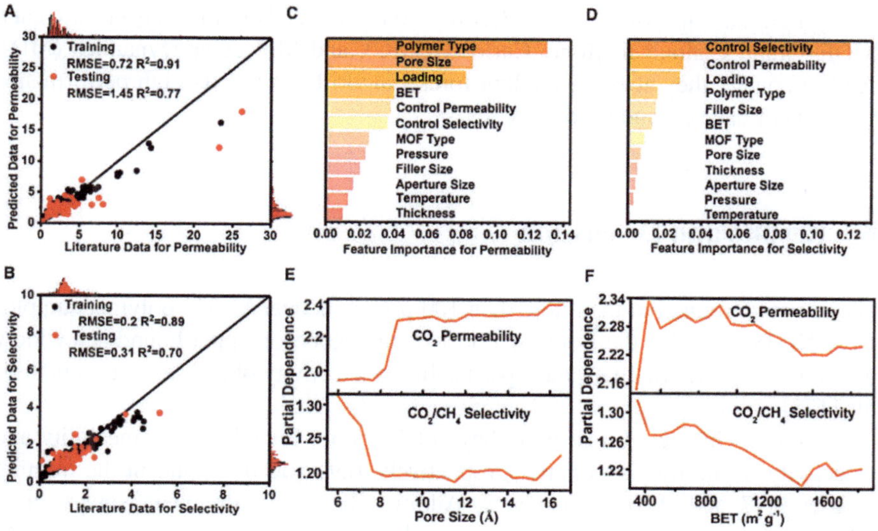


**Fig. 7** Machine learning understanding of MOF-based MMMs for $CO_2/CH_4$ separation based on literature data (**a–f**) The scatterplots showing distributions of predicted and literature values of (**a**) relative CO permeability and (**b**) relative $CO_2/CH_4$ selectivity in the training and testing datasets; feature importance ranking for (**c**) relative CO permeability and (**d**) relative $CO_2/CH_4$ selectivity; partial dependence plots of permeability and selectivity on (**e**) pore size and (**f**) BET surface area

matches well with optimal design analyzed by ML in terms of pore size and surface area, higher than Cu-THQ. These MOFs particles are blended with polymers, Pebax 2533 and PIM-1. Finally, they tried to transfer learning the data obtained from $CO_2/CH_4$ separation to $CO_2/N_2$ separation. They extracted 398 datasets on MOF-based MMMs for $CO_2/N_2$ separation, and used a transfer learning method with random forests since both applications have similarity. The datasets are used to train another model, with the predictive output as direct prediction and the output transferred from $CO_2/CH_4$ as transfer learning prediction.

To summarize their work methods, first, they collected data and description. Then, they use ML algorithms to pre-processing the datasets. They use several algorithms, including KNN algorithm for imputation, random forests (and transfer learning), and partial dependence computing. Finally, they synthesized the Cu-CAT-1 and Cu-THQ, prepared the mixed matrix membranes, and tested single gas permeation.

Hilal Daglar and Seda Keskin did similar works. They used molecular simulations and machine learning to investigate the gas separation performances of MOF membranes and MOF/polymer MMMs for $He/H_2$, $H3/N_2$, $He/CH_4$, $H_2/N_2$, $H_2/CH_4$, and $N_2/CH_4$. They developed eight different ML algorithms, including LassoLarsCV, ETR, GBR, and RFR algorithms to predict the uptake and diffusivity of the gases. Their main goal is to develop ML models that are useful to evaluate the gas separation performances of a variety of MOFs and MOF/polymer membranes.

Z. Zhang et al. use the ML to help high-throughput prediction of ionic liquid at MOF composites for carbon capture. They use simulation methods, such as configurational-bias Monte Carlo (CBMC) to insert ([$NH_2$-Pmim][$Tf_2N$]) into each MOF, Grand Canonical Monte Carlo (GCMC), and Molecular Dynamics (MD). And of course, they also use random forests algorithm since its high performance with lots of datasets and high accuracy.

## 6 Results and Discussion

Barnett et al. found out that more than 100 polymers over 11,325 polymer samples from NIMS exceeded the Robeson upper bound. Robeson upper bound is a widely acknowledged criteria that state a permeability-selectivity of gas separation materials [173].

Based on Guan et al. research, they got $R^2$ values for relative permeability and selectivity in the training dataset, respectively, 0.91 and 0.89; and in the training dataset 0.77 and 0.70, respectively. Additionally, low root mean square error (RMSE) values and well-matches between the prediction and the literature data. By temporarily removing mixing input values, it resulted in close $R^2$, RMSE, and mean absolute error values between test and training dataset. For an optimized performance of MMMs, they found out that pore size of ~12 Angstrom or higher is the best, and a medium BET value of `800 $m^2$ $g^{-1}$ is good for balanced membrane performances. These results and models showed us that incorporation of MOFs into MMMs has a possibility to mitigate the plasticization phenomenon. For the MOFs-polymers blend, Cu-CAT-1-Pebax 2533 produced higher $CO_2$ permeability than Cu-THQ, even better than other MOFs-Pebax. The resilience of MMM to plasticization can be improved by increasing $CO_2$ molar concentration from 20% to 80%. Finally, for the transfer learning to $CO_2/N_2$, the result showed a better performance than direct prediction. These results validate the potential of transfer learning in the field of carbon-capturing. Guan et al. showed their results in Fig. 8.

Hilal Daglar and Seda Keskin research showed a high accuracy of ML models for gas uptake and diffusion properties of MOFs. It is revealed that physical properties of MOFs are more critical than chemical descriptors for predicting gas adsorption and diffusion. Their ML models are useful to estimate gas separation performances of MOF-based membranes in a rapid and accurate manner.

Based on the literature review conducted, carbon-based membranes have shown significant potential in water and gas separation applications, particularly in desalination, wastewater treatment, and industrial gas purification. Recent studies demonstrate that the incorporation of materials such as graphene oxide (GO), carbon nanotubes (CNT), and carbon quantum dots (CQDs) can significantly enhance membrane performance in terms of permeability, selectivity, and fouling resistance.

In the context of desalination, freestanding membranes such as CNT-based buckypaper and stacked GO layers have demonstrated high salt rejection efficiency and increased water flux. However, technical challenges remain, including limited mechanical stability, layer swelling, and difficulties in large-scale production. On the other hand, composite membranes that combine carbon nanomaterials with

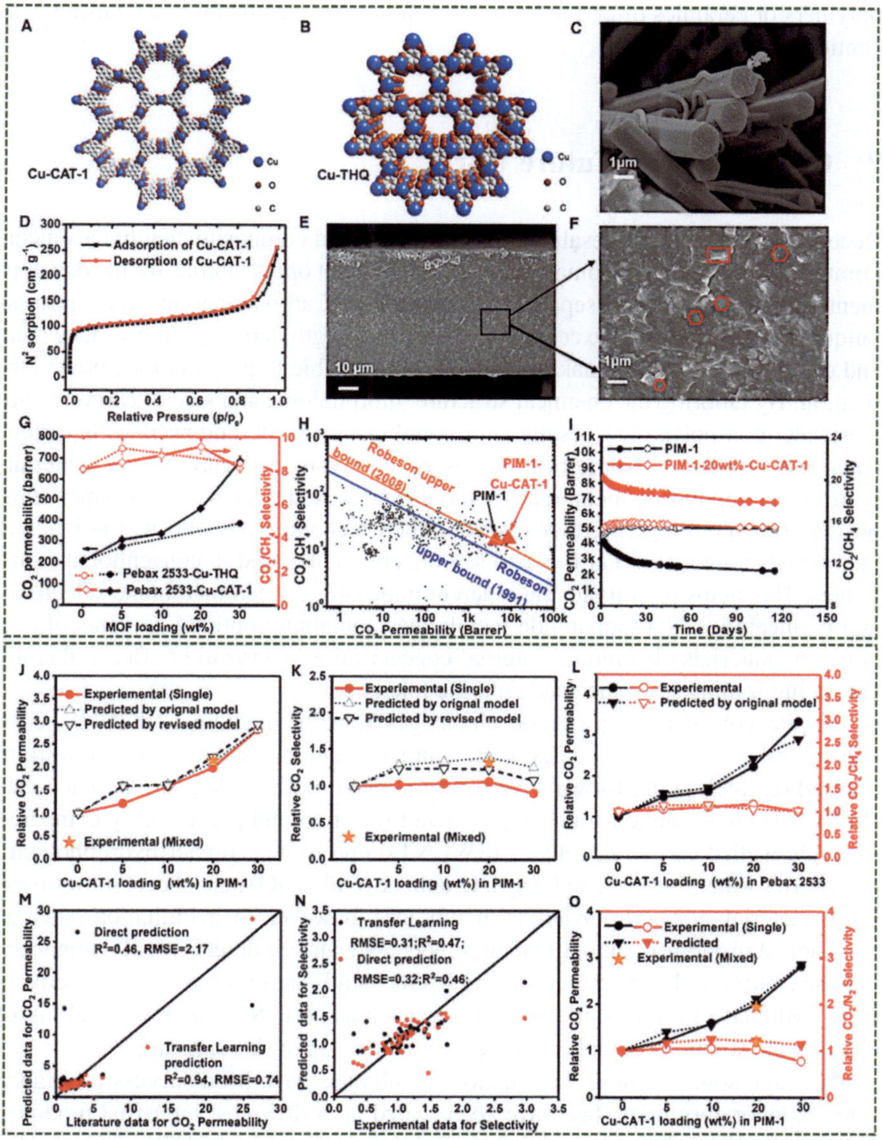


**Fig. 8** Characterization of Cu-CAT-1, Cu-THQ, and Mixed Matrix Membranes (MMMs). (**A–B**) Schematics of Cu-CAT-1 and Cu-THQ structures; (**C**) SEM image of Cu-CAT-1; (**D**) Nitrogen adsorption-desorption isotherms; (**E–F**) Cross-sectional SEM images of Pebax/Cu-CAT-1 (20 wt%) membranes; (**G**) $CO_2$ permeability and selectivity of Pebax membranes; (**H**) Performance comparison of PIM-1 membranes against literature data (Robeson plot); (**I**) Long-term stability test of permeability and selectivity. Experimental vs. Predicted Gas Separation Performance of Cu-CAT-1 MMMs. (**J–K**) Relative $CO_2$ permeability and $CO_2/CH_4$ selectivity for PIM-1-based membranes; (**L**) Performance metrics for Pebax 2533-based membranes; (**M–N**) Comparison of direct vs. transfer learning models based on literature data; (**O**) Transfer learning predictions for PIM-1 membranes. Yellow stars indicate mixed gas test data; all other data are from pure gas tests

polymers or ceramics offer a more stable solution and can be fabricated using conventional techniques.

## 7 Conclusion and Future Work

Recent advancements in desalination membranes and composite membranes incorporating carbon-based nanomaterials offer significant opportunities for the development of advanced water separation membranes. Carbon nanomaterials possess unique properties such as excellent mechanical strength, large specific surface area, and ease of modification, making them highly valuable in polymer membrane fabrication. By tailoring the chemical structure, morphology, and size of carbon nanomaterials, non-polyamide desalination membranes can be endowed with various enhanced properties. Numerous studies have investigated carbon nanomaterials both as self-supporting membrane materials and as additives within polymer composites. Incorporating single or hybrid carbon nanomaterials into non-polyamide membranes improves water flux and salt rejection compared to conventional membranes. This helps to mitigate the inherent trade-off between membrane selectivity and permeability. Non-polyamide membranes containing optimal amounts of carbon nanomaterials also reduce internal concentration polarization effects, thereby enhancing fouling resistance and chlorine tolerance.

Despite considerable progress, challenges remain, including reliability, cost, regenerability, safety, and large-scale implementation. Several aspects warrant focused research going forward. Vertically aligned multi-walled carbon nanotubes (MWCNTs) have demonstrated superior salt rejection and permeability compared to single-walled carbon nanotubes (SWCNTs) due to their higher hydrophobicity and shorter water transport pathways. Graphene oxide (GO) nanosheets show promise as materials for non-polyamide membranes; however, their limitations such as mechanical instability, suboptimal alignment, and surface defects require attention. Precise control of the interlayer spacing in GO membranes is essential to effectively sieve hydrated salt ions while allowing water transport. Nevertheless, swelling of GO in ionic solutions complicates accurate regulation of this spacing.

Although some of these desalination membranes have been successfully fabricated at laboratory scale, few have been scaled up to low-cost industrial modules, resulting in slow commercial adoption. High synthesis costs of CNTs and graphene, coupled with concerns over reproducibility and feasibility of membrane fabrication methods, impede commercialization. Moreover, technologies for producing mechanically robust membranes remain immature. The price of monolayer graphene remains substantially higher than that of commercial reverse osmosis membranes.

Additionally, the limited structural stability of nanocomposite membranes can lead to the leaching of carbon nanomaterials into the environment during operation. This phenomenon not only reduces membrane performance and lifespan but also raises environmental concerns due to potential toxicity to humans, animals, and ecosystems. Enhancing carbon membrane synthesis methods to ensure strong binding of nanomaterials—through covalent bonding, electrostatic interactions, or similar strategies—is imperative. Furthermore, implementation of monitoring strategies for nanomaterial leaching and comprehensive environmental and human toxicity risk assessments is essential. Overall, carbon-based nanomaterials have emerged as pioneers in the advancement of desalination membranes for water purification.

## References


1. Kılıç, Z.: The importance of water and conscious use of water. Int. J. Hydrol. **4**(5), 239–241 (2020)
2. Jackson, R.B., et al.: Water in a changing world. Ecol. Appl. **11**(4), 1027–1045 (2001)
3. He, C., et al.: Future global urban water scarcity and potential solutions. Nat. Commun. **12**(1), 4667 (2021)
4. Mishra, R.K.: Fresh water availability and its global challenge. Br. J. Multidiscip. Adv Stud. **4**(3), 1–78 (2023)
5. Bjelić, L.S., et al.: Characteristics of saline water and health impact. In: Removal of Pollutants from Saline Water, pp. 7–28. CRC Press (2021)
6. Darre, N.C., Toor, G.S.: Desalination of water: a review. Curr. Pollut. Rep. **4**(2), 104–111 (2018)
7. Elimelech, M., Phillip, W.A.: The future of seawater desalination: energy, technology, and the environment. Science. **333**(6043), 712–717 (2011)
8. Sinha, S., et al.: Nanotechnology-based solutions for antibiofouling applications: an overview. ACS Appl. Nano Mater. **6**(14), 12828–12848 (2023)
9. Song, Y., et al.: Porous materials for water purification. Angew. Chem. **135**(11), e202216724 (2023)
10. Austria, H.F.M., et al.: Tailoring the specific crosslinking sites of graphene oxide framework nanosheets for controlled nanofiltration of salts and dyes. J. Clean. Prod. **395**, 136280 (2023)
11. Subrahmanya, T., et al.: An eco-friendly and reusable syringe filter membrane for the efficient removal of dyes from water via low pressure filtration assisted self-assembling of graphene oxide and SBA-15/PDA. J. Clean. Prod. **349**, 131425 (2022)
12. Austria, H.F.M., et al.: Investigation of the pH-mediated fabrication process of pure and polyethyleneimine-crosslinked graphene oxide membranes for desalination and heavy metal ion separation. Carbon. **224**, 119019 (2024)
13. Li, J.-R., Kuppler, R.J., Zhou, H.-C.: Selective gas adsorption and separation in metal–organic frameworks. Chem. Soc. Rev. **38**(5), 1477–1504 (2009)
14. Widakdo, J., et al.: Graphene oxide-melamine nanofilm composite membrane for efficient $CO_2$ gas separation. Sep. Purif. Technol. **323**, 124521 (2023)
15. Widakdo, J., et al.: Bioinspired ionic liquid-graphene based smart membranes with electrical tunable channels for gas separation. Appl. Mater. Today. **27**, 101441 (2022)
16. Widakdo, J., et al.: Mechanism of a self-assembling smart and electrically responsive PVDF–graphene membrane for controlled gas separation. ACS Appl. Mater. Interfaces. **12**(27), 30915–30924 (2020)
17. Widakdo, J., et al.: Switching gas permeation through smart membranes by external stimuli: a review. J. Mater. Chem. A. **10**(32), 16743–16760 (2022)
18. Widakdo, J., et al.: Tailoring of graphene–organic frameworks membrane to enable reversed electrical-switchable permselectivity in $CO_2$ separation. Carbon. **182**, 545–558 (2021)
19. Marchetti, P., Peeva, L., Livingston, A.: The selectivity challenge in organic solvent nanofiltration: membrane and process solutions. Annu. Rev. Chem. Biomol. Eng. **8**(1), 473–497 (2017)

20. Subrahmanya, T., et al.: Effect of functionalized nanodiamonds and surfactants mediation on the nanofiltration performance of polyamide thin-film nanocomposite membranes. Desalination. **555**, 116540 (2023)
21. Widakdo, J., et al.: Positron annihilation spectroscopy for the free volume depth profile analysis of multilayer and 2D materials composite membranes: A review. Sep. Purif. Technol. **322**, 124366 (2023)
22. Abebe, S.H., et al.: Zwitterionic M-PhA modified GO nanocomposite membranes for enhanced organic solvent nanofiltration and wastewater treatment with antifouling performance. J. Membr. Sci., 124596 (2025)
23. Holst, J.R., Cooper, A.I.: Ultrahigh Surface Area in Porous Solids. Wiley Online Library (2010)
24. Zhai, W., et al.: Recent progress on wear-resistant materials: designs, properties, and applications. Adv. Sci. **8**(11), 2003739 (2021)
25. Wu, J., et al.: Porous polymers as multifunctional material platforms toward task-specific applications. Adv. Mater. **31**(4), 1802922 (2019)
26. Li, C., et al.: Carbon-based membrane materials and applications in water and wastewater treatment: A review. Environ. Chem. Lett. **19**(2), 1457–1475 (2021)
27. Roy, S., Ragunath, S.: Emerging membrane technologies for water and energy sustainability: Future prospects, constraints and challenges. Energies. **11**(11), 2997 (2018)
28. Mauter, M.S., Elimelech, M.: Environmental applications of carbon-based nanomaterials. Environ. Sci. Technol. **42**(16), 5843–5859 (2008)
29. Regmi, C., Kshetri, Y.K., Wickramasinghe, S.R.: Carbon-based nanocomposite membranes for membrane distillation: progress, problems and future prospects. Membranes. **14**(7), 160 (2024)
30. Khademolqorani, S., et al.: Innovative integration of layered carbon materials in biopolymer fibrous membranes for sustainable water treatment. Adv. Sustain. Syst. **9**, 2500035 (2025)
31. Gaudio, M.T., et al.: Artificial intelligence-based optimization of industrial membrane processes. Earth Syst. Environ. **5**(2), 385–398 (2021)
32. Saha, P., et al.: Hybrid membrane process for water treatment: a short review. Aqua Water Infrastruct. Ecosys. **72**(5), 608–622 (2023)
33. Khulbe, K., Matsuura, T.: Removal of heavy metals and pollutants by membrane adsorption techniques. Appl Water Sci. **8**(1), 19 (2018)
34. Ang, W.L., et al.: A review on the applicability of integrated/hybrid membrane processes in water treatment and desalination plants. Desalination. **363**, 2–18 (2015)
35. Sun, S., Kumar, A., Yen, J.: Merging workflows: A new perspective on connecting business processes. Decis. Support. Syst. **42**(2), 844–858 (2006)
36. Judeinstein, P., Sanchez, C.: Hybrid organic–inorganic materials: a land of multidisciplinarity. J. Mater. Chem. **6**(4), 511–525 (1996)
37. Sinha Ray, S., et al.: Recent developments in nanomaterials-modified membranes for improved membrane distillation performance. Membranes. **10**(7), 140 (2020)
38. Meot-Ner, M.: The ionic hydrogen bond. Chem. Rev. **105**(1), 213–284 (2005)
39. Peng, F., et al.: Hybrid organic– inorganic membrane: solving the tradeoff between permeability and selectivity. Chem. Mater. **17**(26), 6790–6796 (2005)
40. Li, C., et al.: Developments of carbon-based membrane materials for water treatment. In: Membranes for Environmental Applications, pp. 121–175. Springer (2020)
41. Khan, Z.U., Kausar, A., Ullah, H.: A review on composite papers of graphene oxide, carbon nanotube, polymer/GO, and polymer/CNT: Processing strategies, properties, and relevance. Polym.-Plast. Technol. Eng. **55**(6), 559–581 (2016)
42. Liu, X.-W., et al.: Composites of metal–organic frameworks and carbon-based materials: preparations, functionalities and applications. J. Mater. Chem. A. **4**(10), 3584–3616 (2016)
43. Qian, Q., et al.: MOF-based membranes for gas separations. Chem. Rev. **120**(16), 8161–8266 (2020)
44. Satyam, S., Patra, S.: Innovations and challenges in adsorption-based wastewater remediation: A comprehensive review. Heliyon. **10**(9), e29573 (2024)

45. Akhtar, M.S., Ali, S., Zaman, W.: Innovative adsorbents for pollutant removal: exploring the latest research and applications. Molecules. **29**(18), 4317 (2024)
46. Iravani, A., Akbari, M.H., Zohoori, M.: Advantages and disadvantages of green technology; goals, challenges and strengths. Int. J. Sci. Eng. Appl. **6**(9), 272–284 (2017)
47. Díez-Pascual, A.M.: Carbon-based nanomaterials. p. 7726 (2021)
48. Kaur, I.P., et al.: Issues and concerns in nanotech product development and its commercialization. J. Control. Release. **193**, 51–62 (2014)
49. Zarghami, S., Mohammadi, T.: Recent advances on carbon-based nanocomposite membranes in water and wastewater applications. Curr. Opin. Green Sustain. Chem. **53**, 101026 (2025)
50. Kumar, R., Singh, R.K., Singh, D.P.: Natural and waste hydrocarbon precursors for the synthesis of carbon based nanomaterials: graphene and CNTs. Renew. Sust. Energ. Rev. **58**, 976–1006 (2016)
51. Fagiolari, L., Bella, F.: Carbon-based materials for stable, cheaper and large-scale processable perovskite solar cells. Energy Environ. Sci. **12**(12), 3437–3472 (2019)
52. Bramsiepe, C., et al.: Low-cost small scale processing technologies for production applications in various environments—Mass produced factories. Chem. Eng. Process. Process Intensif. **51**, 32–52 (2012)
53. Shi, Y., et al.: Performance stability and regeneration property of catalytic membranes coupled with advanced oxidation process: A comprehensive review. Sustainability. **15**(9), 7556 (2023)
54. Zhou, W., et al.: Electrochemical regeneration of carbon-based adsorbents: a review of regeneration mechanisms, reactors, and future prospects. Chem. Eng. J. Adv. **5**, 100083 (2021)
55. Aslam, M.M.-A., et al.: Functionalized carbon nanotubes (CNTs) for water and wastewater treatment: preparation to application. Sustainability. **13**(10), 5717 (2021)
56. Haider, T.P., et al.: Plastics of the future? The impact of biodegradable polymers on the environment and on society. Angew. Chem. Int. Ed. **58**(1), 50–62 (2019)
57. Soete, L.: International diffusion of technology, industrial development and technological leapfrogging. World Dev. **13**(3), 409–422 (1985)
58. Widakdo, J., et al.: Dual functional GO-Ag incorporated nanocomposite pervaporation membrane with alcohol dehydration performance and enhanced antibacterial property. Mater. Today Chem. **24**, 100985 (2022)
59. Shananda, N., et al.: Carbon-based nano photocatalyst with superior chemical and physical properties in enhancing hydrogen evolution reaction. In: Photocatalytic Hydrogen Fuel Generation: Designing Highly Efficient Semiconductor Materials, pp. 205–249. Springer (2025)
60. Sivakumar, M., et al.: Porous graphene nanoplatelets encompassed with nitrogen and sulfur group for heavy metal ions removal of adsorption and desorption from single or mixed aqueous solution. Sep. Purif. Technol. **288**, 120485 (2022)
61. Sianipar, M., et al.: Functionalized carbon nanotube (CNT) membrane: progress and challenges. RSC Adv. **7**(81), 51175–51198 (2017)
62. Lee, J., Jeong, S., Liu, Z.: Progress and challenges of carbon nanotube membrane in water treatment. Crit. Rev. Environ. Sci. Technol. **46**(11-12), 999–1046 (2016)
63. Liu, D., et al.: Chemical approaches to carbon-based metal-free catalysts. Adv. Mater. **31**(13), 1804863 (2019)
64. Liu, W.-J., Jiang, H., Yu, H.-Q.: Development of biochar-based functional materials: toward a sustainable platform carbon material. Chem. Rev. **115**(22), 12251–12285 (2015)
65. Sun, X., et al.: Advance in using plasma technology for modification or fabrication of carbon-based materials and their applications in environmental, material, and energy fields. Adv. Funct. Mater. **31**(7), 2006287 (2021)
66. Park, H.B., et al.: Maximizing the right stuff: The trade-off between membrane permeability and selectivity. Science. **356**(6343), eaab0530 (2017)
67. Werber, J.R., Deshmukh, A., Elimelech, M.: The critical need for increased selectivity, not increased water permeability, for desalination membranes. Environ. Sci. Technol. Lett. **3**(4), 112–120 (2016)

68. Moustafa, H., Isawi, H., Abd El Wahab, S.: Utilization of PVA nano-membrane based synthesized magnetic GO-Ni-$Fe_2O_4$ nanoparticles for removal of heavy metals from water resources. Environ. Nanotechnol. Monit. Manag. **18**, 100696 (2022)
69. Cai, Z., et al.: Application of nanotechnologies for removing pharmaceutically active compounds from water: development and future trends. Environ. Sci. Nano. **5**(1), 27–47 (2018)
70. Gao, G., Pan, M., Vecitis, C.D.: Effect of the oxidation approach on carbon nanotube surface functional groups and electrooxidative filtration performance. J. Mater. Chem. A. **3**(14), 7575–7582 (2015)
71. Yu, W., et al.: Progress in the functional modification of graphene/graphene oxide: A review. RSC Adv. **10**(26), 15328–15345 (2020)
72. Rana, D., Matsuura, T.: Surface modifications for antifouling membranes. Chem. Rev. **110**(4), 2448–2471 (2010)
73. Soltys, L., et al.: Green synthesis of metal and metal oxide nanoparticles: Principles of green chemistry and raw materials. Magnetochemistry. **7**(11), 145 (2021)
74. Panhwar, S., et al.: Importance and analytical perspective of green synthetic strategies of copper, zinc, and titanium oxide nanoparticles and their applications in pathogens and environmental remediation. Curr. Anal. Chem. **17**(8), 1169–1181 (2021)
75. Smijs, T.G., Pavel, S.: Titanium dioxide and zinc oxide nanoparticles in sunscreens: focus on their safety and effectiveness. Nanotechnol. Sci. Appl., 95–112 (2011)
76. Bora, P., et al.: Carbon nanomaterials for designing next-generation membranes and their emerging applications. Chem. Commun. **59**(76), 11320–11336 (2023)
77. Ridgway, H.F., Orbell, J., Gray, S.: Molecular simulations of polyamide membrane materials used in desalination and water reuse applications: Recent developments and future prospects. J. Membr. Sci. **524**, 436–448 (2017)
78. Thakur, V.K., Thakur, M.K.: Chemical Functionalization of Carbon Nanomaterials. CRC Press Warentown (NJ) (2018)
79. Sabir, A., et al.: Carbon nanotube and graphene oxide based membranes. In: Nanoscale Materials in Water Purification, pp. 361–381. Elsevier (2019)
80. Shao, D.-D., et al.: Amine–carbon quantum dots (CQDs–$NH_2$) tailored polymeric loose nanofiltration membrane for precise molecular separation. Chem. Eng. Res. Des. **171**, 237–246 (2021)
81. Köhler, M.H., et al.: Three-dimensional and lamellar graphene oxide membranes for water purification. In: Two-Dimensional (2D) Nanomaterials in Separation Science, pp. 87–111. Springer (2021)
82. Sun, P., Wang, K., Zhu, H.: Recent developments in graphene-based membranes: structure, mass-transport mechanism and potential applications. Adv. Mater. **28**(12), 2287–2310 (2016)
83. Hummer, G., Rasaiah, J.C., Noworyta, J.P.: Water conduction through the hydrophobic channel of a carbon nanotube. Nature. **414**(6860), 188–190 (2001)
84. Xu, M., et al.: Graphene-like two-dimensional materials. Chem. Rev. **113**(5), 3766–3798 (2013)
85. Sun, M., Li, J.: Graphene oxide membranes: Functional structures, preparation and environmental applications. Nano Today. **20**, 121–137 (2018)
86. Fathizadeh, M., et al.: Graphene oxide: a novel 2-dimensional material in membrane separation for water purification. Adv. Mater. Interfaces. **4**(5), 1600918 (2017)
87. Lin, T., et al.: Recent advances in graphene-based membranes with nanochannels and nanopores. Small Struct. **6**(1), 2400320 (2025)
88. Qian, Y., et al.: Tuning interlayer spacing of graphene oxide membranes with enhanced desalination performance. Desalination. **460**, 56–63 (2019)
89. Hung, W.-S., et al.: Tuning the interlayer spacing of forward osmosis membranes based on ultrathin graphene oxide to achieve desired performance. Carbon. **142**, 337–345 (2019)
90. Ayappa, K.G.: Enhancing the dynamics of water confined between graphene oxide surfaces with janus interfaces: a molecular dynamics study. J. Phys. Chem. B. **123**(13), 2978–2993 (2019)

91. Kang, E., et al.: Large-scale, ultrapliable, and free-standing nanomembranes. Adv. Mater. **25**(15), 2167–2173 (2013)
92. Shi, Q., Cheng, W.: Free-standing 2D nanoassemblies. Adv. Funct. Mater. **30**(2), 1902301 (2020)
93. Pahal, S., et al.: Nanoarchitectonics for Free-standing polyelectrolyte multilayers films: exploring the flipped surfaces. ChemNanoMat. **9**(2), e202200462 (2023)
94. Werber, J.R., Porter, C.J., Elimelech, M.: A path to ultraselectivity: Support layer properties to maximize performance of biomimetic desalination membranes. Environ. Sci. Technol. **52**(18), 10737–10747 (2018)
95. Liu, Q., et al.: Hydrophilic mineral coating of membrane substrate for reducing internal concentration polarization (ICP) in forward osmosis. Sci. Rep. **6**(1), 1–10 (2016)
96. Phuntsho, S., et al.: Free-standing, thin-film, symmetric membranes: Next-generation membranes for engineered osmosis. J. Membr. Sci. **607**, 118145 (2020)
97. He, Y., et al.: Ultrathin two-dimensional membranes by assembling graphene and MXene nanosheets for high-performance precise separation. J. Mater. Chem. A. **12**(44), 30121–30168 (2024)
98. Sadeghi, I., Kaner, P., Asatekin, A.: Controlling and expanding the selectivity of filtration membranes. Chem. Mater. **30**(21), 7328–7354 (2018)
99. Foorginezhad, S., et al.: Emerging membrane technologies for sustainable water treatment: A review of recent advances. Environ. Sci. Adv. **4**, 530–570 (2025)
100. Yang, Y., et al.: Scalable, flexible, ultra-strong, free-standing papery all carbon nanotube membrane with excellent separation and antifouling properties. Chem. Eng. J. **490**, 151750 (2024)
101. Ribeiro, B., et al.: Carbon nanotube buckypaper reinforced polymer composites: a review. Polímeros. **27**(03), 247–255 (2017)
102. Rashid, M.H.-O.: Synthesis, Characterisation and Water Purification Properties of Carbon Nanotube Membranes (Buckypapers). University of Wollongong (2016)
103. Jiang, Q., et al.: Mechanical, electrical and thermal properties of aligned carbon nanotube/polyimide composites. Compos. Part B. **56**, 408–412 (2014)
104. Deng, L., et al.: The effective Young's modulus of carbon nanotubes in composites. ACS Appl. Mater. Interfaces. **3**(2), 433–440 (2011)
105. Roy, S., et al.: Formation of carbon nanotube bucky paper and feasibility study for filtration at the nano and molecular scale. J. Phys. Chem. C. **116**(35), 19025–19031 (2012)
106. Rashid, M.H.-O., Ralph, S.F.: Carbon nanotube membranes: synthesis, properties, and future filtration applications. Nanomaterials. **7**(5), 99 (2017)
107. Alsuhybani, M., Alshahrani, A., Haidyrah, A.S.: Synthesis, characterization, and evaluation of evaporated casting MWCNT/chitosan composite membranes for water desalination. J. Chem. **2020**(1), 5207680 (2020)
108. Dumée, L.F., et al.: Characterization and evaluation of carbon nanotube Bucky-Paper membranes for direct contact membrane distillation. J. Membr. Sci. **351**(1-2), 36–43 (2010)
109. Yang, Y., et al.: Large-area graphene-nanomesh/carbon-nanotube hybrid membranes for ionic and molecular nanofiltration. Science. **364**(6445), 1057–1062 (2019)
110. Saeed, G., et al.: Two-dimensional (2D) material nanofiltration membranes for effective recovery of lithium. J. Ind. Eng. Chem. **150**, 116–133 (2025)
111. Rollings, R.C., Kuan, A.T., Golovchenko, J.A.: Ion selectivity of graphene nanopores. Nat. Commun. **7**(1), 11408 (2016)
112. Ge, R., et al.: GO-based membranes for desalination. Membranes. **13**(2), 220 (2023)
113. Fan, L., Zheng, J.: Preparation, types, and applications of one-and two-dimensional nanochannels and their transport properties for water and ions. Nanotechnol. Rev. **13**(1), 20240097 (2024)
114. Alkhouzaam, A., Qiblawey, H.: Functional GO-based membranes for water treatment and desalination: Fabrication methods, performance and advantages. A review. Chemosphere. **274**, 129853 (2021)

115. Zhang, H., Li, X., Xu, T.: Two-dimensional graphene oxide nanochannel membranes for ionic separation. Curr. Opin. Chem. Eng. **39**, 100899 (2023)
116. Wang, Y., et al.: Recent progress on carbon based desalination membranes and carbon nanomaterial incorporated non-polyamide desalination membranes. J. Environ. Chem. Eng. **9**(4), 105762 (2021)
117. Li, Y., et al.: Recent advances in the fabrication of advanced composite membranes. J. Mater. Chem. A. **1**(35), 10058–10077 (2013)
118. Bosco, C.D., et al.: Carbon nanomaterial-based membranes in solid-phase extraction. Microchim. Acta. **190**(5), 175 (2023)
119. Koros, W., Coleman, M., Walker, D.: Controlled permeability polymer membranes. Annu. Rev. Mater. Sci. **22**(1), 47–89 (1992)
120. Joshi, U.A., Sharma, S.C., Harsha, S.P.: Effect of carbon nanotube orientation on the mechanical properties of nanocomposites. Compos. Part B. **43**(4), 2063–2071 (2012)
121. Ahmed, M.A., Mahmoud, S.A., Mohamed, A.A.: Nanomaterials-modified reverse osmosis membranes: a comprehensive review. RSC Adv. **14**(27), 18879–18906 (2024)
122. Gittleman, C.S., Coms, F.D., Lai, Y.-H.: Membrane durability: physical and chemical degradation. In: Polymer electrolyte Fuel Cell Degradation, p. 15 (2011)
123. Al Harby, N.F., El-Batouti, M., Elewa, M.M.: Prospects of polymeric nanocomposite membranes for water purification and scalability and their health and environmental impacts: A review. Nano. **12**(20), 3637 (2022)
124. Lau, W.J., et al.: A recent progress in thin film composite membrane: A review. Desalination. **287**, 190–199 (2012)
125. Lawler, J.: Incorporation of graphene-related carbon nanosheets in membrane fabrication for water treatment: A review. Membranes. **6**(4), 57 (2016)
126. Cheng, Y., et al.: Advanced porous materials in mixed matrix membranes. Adv. Mater. **30**(47), 1802401 (2018)
127. Ismail, A.F., Khulbe, K.C., Matsuura, T.: Membrane fabrication/manufacturing techniques. In: Gas Separation Membranes: Polymeric and Inorganic, pp. 193–220. Springer (2015)
128. Hofs, B., et al.: Comparison of ceramic and polymeric membrane permeability and fouling using surface water. Sep. Purif. Technol. **79**(3), 365–374 (2011)
129. Dong, Y., et al.: Stable superhydrophobic ceramic-based carbon nanotube composite desalination membranes. Nano Lett. **18**(9), 5514–5521 (2018)
130. Yuan, Z., et al.: Carbon composite membranes for thermal-driven membrane processes. Carbon. **179**, 600–626 (2021)
131. Van der Bruggen, B., et al.: A review of pressure-driven membrane processes in wastewater treatment and drinking water production. Environ. Prog. **22**(1), 46–56 (2003)
132. Charcosset, C.: Ultrafiltration, microfiltration, nanofiltration and reverse osmosis in integrated membrane processes. In: Integrated Membrane Systems and Processes, pp. 1–22 (2016)
133. Liao, Y., et al.: Progress in electrospun polymeric nanofibrous membranes for water treatment: Fabrication, modification and applications. Prog. Polym. Sci. **77**, 69–94 (2018)
134. Fritzmann, C., et al.: State-of-the-art of reverse osmosis desalination. Desalination. **216**(1-3), 1–76 (2007)
135. Sabir, A., et al.: Fabrication of tethered carbon nanotubes in cellulose acetate/polyethylene glycol-400 composite membranes for reverse osmosis. Carbohydr. Polym. **132**, 589–597 (2015)
136. Khan, A.A., et al.: Metal oxide and carbon nanomaterial based membranes for reverse osmosis and membrane distillation: A comparative review. Environ. Res. **202**, 111716 (2021)
137. Ortiz-Medina, J., et al.: Nanostructured carbon-based membranes: nitrogen doping effects on reverse osmosis performance. NPG Asia Mater. **8**(4), e258–e258 (2016)
138. Ghaseminezhad, S.M., Barikani, M., Salehirad, M.: Development of graphene oxide-cellulose acetate nanocomposite reverse osmosis membrane for seawater desalination. Compos. Part B. **161**, 320–327 (2019)

139. Falath, W., Sabir, A., Jacob, K.I.: Highly improved reverse osmosis performance of novel PVA/DGEBA cross-linked membranes by incorporation of Pluronic F-127 and MWCNTs for water desalination. Desalination. **397**, 53–66 (2016)
140. Zhang, Y., et al.: A high-flux and anti-interference dual-functional membrane for effective removal of Pb (II) from natural water. J. Hazard. Mater. **384**, 121492 (2020)
141. Chen, W., et al.: High-flux water desalination with interfacial salt sieving effect in nanoporous carbon composite membranes. Nat. Nanotechnol. **13**(4), 345–350 (2018)
142. Yang, Z., et al.: A review on reverse osmosis and nanofiltration membranes for water purification. Polymers. **11**(8), 1252 (2019)
143. Tian, X., et al.: Enhancing membrane materials for efficient Li recycling and recovery. Adv. Mater. **37**(5), 2402335 (2025)
144. Nghiem, L.D., Hawkes, S.: Effects of membrane fouling on the nanofiltration of trace organic contaminants. Desalination. **236**(1-3), 273–281 (2009)
145. Sun, W., et al.: Pretreatment and membrane hydrophilic modification to reduce membrane fouling. Membranes. **3**(3), 226–241 (2013)
146. Shen, L., et al.: Polymeric membranes incorporated with ZnO nanoparticles for membrane fouling mitigation: A brief review. Front. Chem. **8**, 224 (2020)
147. Mousa, H.M., et al.: Membranes for oil/water separation: a review. Adv. Mater. Interfaces. **9**(27), 2200557 (2022)
148. Bano, S., et al.: Graphene oxide modified polyamide nanofiltration membrane with improved flux and antifouling properties. J. Mater. Chem. A. **3**(5), 2065–2071 (2015)
149. Hu, R., et al.: Graphene oxide-embedded polyamide nanofiltration membranes for selective ion separation. J. Mater. Chem. A. **5**(48), 25632–25640 (2017)
150. Wang, Z., et al.: High flux nanofiltration membranes with double-walled carbon nanotube (DWCNT) as the interlayer. Membranes. **12**(10), 1011 (2022)
151. Chen, X., et al.: Reduced holey graphene oxide membranes for desalination with improved water permeance. ACS Appl. Mater. Interfaces. **12**(1), 1387–1394 (2019)
152. Gholami, S., et al.: Fabrication of thin-film nanocomposite nanofiltration membranes incorporated with aromatic amine-functionalized multiwalled carbon nanotubes. Rejection performance of inorganic pollutants from groundwater with improved acid and chlorine resistance. Chem. Eng. J. **384**, 123348 (2020)
153. Liu, Y., et al.: Porous organic polymer embedded thin-film nanocomposite membranes for enhanced nanofiltration performance. J. Membr. Sci. **602**, 117982 (2020)
154. Vargas-Figueroa, C., et al.: In-Situ Modification of nanofiltration membranes using carbon nanotubes for water treatment. Membranes. **13**(7), 616 (2023)
155. Li, Z., et al.: Graphene oxide incorporated forward osmosis membranes with enhanced desalination performance and chlorine resistance. Front. Chem. **7**, 877 (2020)
156. Ahmed, D.F., et al.: Graphene oxide incorporated cellulose triacetate/cellulose acetate nanocomposite membranes for forward osmosis desalination. Arab. J. Chem. **14**(3), 102995 (2021)
157. Kwon, S.-B., et al.: High-performance support-free molecular layer-by-layer assembled forward osmosis membrane incorporated with graphene oxide. J. Membr. Sci. **689**, 122152 (2024)
158. Lakra, R., Balakrishnan, M., Basu, S.: Activated carbon incorporation on forward osmosis membrane surface for enhanced performance. Water Supply. **22**(1), 409–423 (2022)
159. Dai, C., et al.: Impact of graphene oxide on properties and structure of thin-film composite forward osmosis membranes. Polymers. **14**(18), 3874 (2022)
160. Humoud, M.S., Roy, S., Mitra, S.: Enhanced performance of carbon nanotube immobilized membrane for the treatment of high salinity produced water via direct contact membrane distillation. Membranes. **10**(11), 325 (2020)
161. Ting, L.L.H., et al.: Development and optimization of low surface free energy of rGO-PVDF mixed matrix membrane for membrane distillation. Sep. Purif. Technol. **305**, 122428 (2023)
162. Leaper, S., et al.: POSS-functionalized graphene oxide/PVDF electrospun membranes for complete arsenic removal using membrane distillation. ACS Appl. Polym. Mater. **3**(4), 1854–1865 (2021)

163. El-Zanati, E., et al.: Functionalized CNT/PVDF blended flat sheet membrane for water desalination via vacuum membrane distillation, fabrication and characterization. Wat. Ener. Food. Env. J. **3**, 53 (2022)
164. Humoud, M.S., Roy, S., Mitra, S.: Scaling reduction in carbon nanotube-immobilized membrane during membrane distillation. Water. **11**(12), 2588 (2019)
165. Grasso, G., et al.: Development of graphene-PVDF composite membranes for membrane distillation. J. Membr. Sci. **604**, 118017 (2020)
166. Yang, G., et al.: Enhanced desalination performance of poly (vinyl alcohol)/carbon nanotube composite pervaporation membranes via interfacial engineering. J. Membr. Sci. **579**, 40–51 (2019)
167. Gupta, O., et al.: Graphene oxide-carbon nanotube (GO-CNT) hybrid mixed matrix membrane for pervaporative dehydration of ethanol. Membranes. **12**(12), 1227 (2022)
168. Dmitrenko, M., et al.: Novel PDMS-b-PPO membranes modified with graphene oxide for efficient pervaporation ethanol dehydration. Membranes. **12**(9), 832 (2022)
169. Zhan, X., et al.: PVA-based MMMs for ethanol dehydration via pervaporation: A comparison study between graphene and graphene oxide. Separations. **9**(2), 26 (2022)
170. Thomas, S., et al.: Multiwalled carbon nanotubes reinforced flexible blend nanocomposites membranes for pervaporation separation of aromatic-aliphatic mixtures. Polym. Polym. Compos. **30**, 09673911211069009 (2022)
171. Zahlan, H., et al.: Separation of benzene/cyclohexane mixtures by pervaporation using poly (ethylene-co-vinylalcohol) and carbon nanotube-filled poly (vinyl alcohol-co-ethylene) membranes. Separations. **7**(4), 68 (2020)
172. Li, S., et al.: Multilevel dispersion induced graphdiyne mixed-matrix membranes for ethanol recovery. Front. Chem. Sci. Eng. **19**(12), 1–10 (2025)
173. Yin, H., et al.: Machine learning for membrane design and discovery. Green Energy Environ. **9**(1), 54–70 (2024)
174. James, G.: An Introduction to Statistical Learning with Applications in R. Springer (2013)
175. Guan, J., et al.: Design and prediction of metal organic framework-based mixed matrix membranes for $CO_2$ capture via machine learning. Cell Rep. Phys. Sci. **3**(0), 100864 (2022)
176. Awad, M, and Rahul K.: Support vector machines for classification. In Efficient learning machines: Theories, concepts, and applications for engineers and system designers, Berkeley, CA: Apress, pp. 39–66. (2015)

# Advancing Carbon-Based Membrane Technologies for the Future

**M. David Domini Oppusunggu, Navilah Auva Hanandara, Nur Halimah, Januar Widakdo, and Hairus Abdullah**

## 1 Introduction

The twenty-first century has amplified the need for sustainable and high-efficiency separation technologies across critical sectors like water management, energy, and chemicals, demanding solutions that outperform conventional, energy-intensive methods such as distillation and cryogenic separation [1, 2]. In this landscape, carbon-based membrane technologies (CMTs) have emerged as a disruptive and highly promising platform [3]. CMTs, which include Carbon Molecular Sieve Membranes (CMSMs) and advanced nanocomposites, offer unique advantages rooted in their precisely controlled microporous structure, exceptional thermal stability, and inherent chemical resistance [4]. This combination often allows them to surpass the trade-off limitations (the Robeson Upper Bound) that restrict traditional polymeric membranes, particularly in harsh operating environments (e.g., high temperatures or aggressive solvents) [5, 6]. The central focus of this chapter is to examine the forward trajectory of CMTs, detailing how they are poised to move from advanced laboratory research to significant future impact. This transition is critically driven by global imperatives, particularly in $CO_2$ mitigation and clean energy separation [7–16]. For instance, CMTs are essential for next-generation $CO_2$ capture [17], natural gas purification ($CO_2/CH_4$ separation) [18], and $H_2$ recovery, offering an energy-efficient alternative to amine scrubbing. Furthermore, carbon nanomaterials, such as Graphene Oxide, are key to developing highly

M. D. D. Oppusunggu · N. A. Hanandara · N. Halimah · J. Widakdo
Department of Physics, Faculty of Mathematics and Natural Sciences, Universitas Indonesia, Depok, Indonesia

H. Abdullah (✉)
Department of Industrial Engineering, Universitas Prima Indonesia, Medan, Indonesia
e-mail: hairus@unprimdn.ac.id



H. Abdullah, J. Widakdo (eds.), *Membrane Physics of Carbon-Based Materials*, Engineering Materials, https://doi.org/10.1007/978-981-95-5605-2_5

fouling-resistant membranes for advanced water treatment processes like Membrane Distillation (MD) [19].

While the scientific opportunities are vast, this chapter will also thoroughly address the necessary preconditions for large-scale success: bridging the gap between high-performance laboratory prototypes and scalable, cost-effective industrial manufacturing [20]. The subsequent sections will therefore outline the promising avenues of material innovation, the role of predictive computational tools, and the significant engineering and economic challenges that must be overcome to realize the commercialization of CMTs.

## 2 Latest Innovations in Carbon Membrane Development

Recent advances in carbon membrane technology are primarily focused on refining the synthesis process and engineering the resulting pore morphology at the subnanometer scale to push performance well beyond the limits of conventional polymers [21]. The innovations can be categorized into three key areas: novel precursor design, advanced synthesis/pyrolysis methods, and the strategic use of carbon-based nanofillers.

### *2.1 Rational Design of Novel Polymeric Precursors*

The foundational performance of any Carbon Molecular Sieve Membrane (CMSM) is intrinsically determined by the chemical structure of its polymer precursor and the subsequent molecular evolution during controlled thermal decomposition, or pyrolysis. This process is not a simple combustion but a highly meticulous thermal 'sculpting' that generates the rigid carbon framework, as broadly illustrated in Fig. 1D. Current research has moved past empirical testing toward rational molecular design, prioritizing precursor chemistries that guarantee the resulting amorphous carbon network possesses the required rigidity and finely tuned ultramicropores [22]. One key strategy exploits polymers that naturally exhibit high free volume, known as Polymers of Intrinsic Microporosity (PIMs). These include structures based on Tröger's Base (TB) or certain rigid-backbone Polyimides (PIs), whose twisted or non-coplanar backbones prevent efficient molecular packing, thus creating large Fractional Free Volume (FFV). Upon pyrolysis, these pre-existing microvoids are refined, yielding the dense network of ultra-micropores (~3–7 Å) critical for the strict molecular sieving mechanism [23]. The success of this strategy is confirmed by the resulting membranes' ability to demonstrate exceptional separation factors, significantly surpassing the limitations of conventional polymers for challenging gas pairs like $H_2/CH_4$. The various structural ideals, including the critical bimodal pore distribution and the ultimate role of ultramicropores in the final CMS framework, are visually represented in Fig. 1C [24].

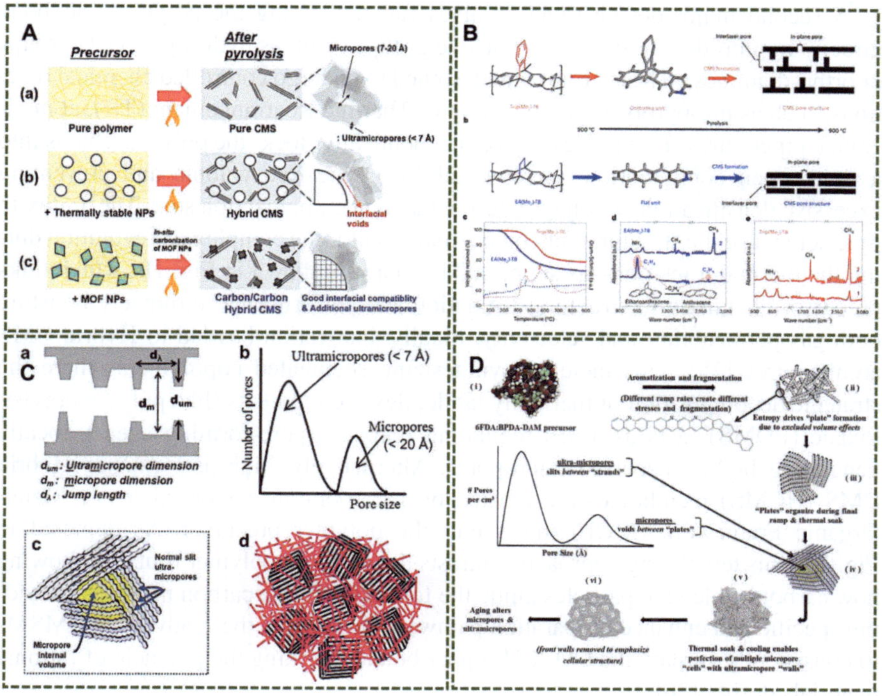


**Fig. 1** (**A**) Schematic illustration of preparation procedure and predicted structure of (**a**) carbon molecular sieve (CMS) membrane derived from pure polymer precursor, (**b**) hybrid CMS (HCMS) membrane derived from mixed matrix membrane (MMM) precursor containing thermally stable inorganic nanoparticles, and (**c**) HCMS membrane derived from MMM precursor containing carbonizable metal-organic framework (MOF) nanoparticles. Note that "NP" in (**b**) and (**c**) stands for "nanoparticle" [26]. Reproduced from ref. [26] with permission from John Wiley and Sons, copyright 2021. (**B**) Preparation and characterization of CMS membranes using 3D ladder polymer precursors (**a, b**), Hypothetical pore structure development for CMS membranes derived from Trip($Me_2$)-TB (**a**) and EA($Me_2$)-TB (**b**) precursors. (**c**), TGA and Gram–Schmidt plots of Trip($Me_2$)-TB and EA($Me_2$)-TB up to 900 °C and under a $N_2$ atmosphere. (**d, e**) Gases detected by TGA–FTIR during the pyrolysis of EA($Me_2$)-TB (**d**) and Trip($Me_2$)-TB (**e**). The spectra of the FTIR labelled by 1 and 2 correspond to the 1 and 2 in the TGA and Gram–Schmidt plots [25]. Reproduced from ref. [25] with permission from Springer Nature, copyright 2023. (**C**) CMS membrane with (**a**) Ideal slit-like pore structure, (**b**) Bimodal distribution of pores, (**c**) idealized micropore cell with ultramicropores between strands in the plate, and (**d**) cellular structure composed with multiple micropore cells with ultramicropore walls surrounded by continuous phase of randomly dispersed strands not included in plate (shown in red to distinguish from strands in plate) believed to reflect the CMS morphology resulting from pyrolysis of conventional random coil polyimide precursors [24]. Reproduced from ref. [24] with permission from John Wiley and Sons, copyright 2021. (**D**) Illustration of the transformation steps of polyimide precursor to amorphous CMS material with micropores and ultramicropores. Reproduced from ref. [22] with permission from Elsevier, copyright 2017

A second major design trend involves functionalizing the polymer backbone directly by introducing thermally reactive groups (such as hydroxyl (–OH) groups or ortho-diamines). These groups are designed to undergo controlled chemical reactions, such as pre-pyrolysis cross-linking or Thermal Rearrangement (TR), at moderate temperatures. This process serves to chemically 'lock' the polymer chains into a rigid, open configuration, thereby stabilizing the free volume and preventing excessive densification or collapse during the final carbonization step. The ability to precisely tailor the structural outcome—and monitor the complex gas evolution during the thermal transition—is supported by analysis shown in Fig. 1B, which links precursor chemistry to structural formation and performance, ultimately boosting both gas permeability and selectivity. Finally, innovation has led to Hybrid CMS architectures [25]. This includes synthesizing segmented copolymers where the strategic decomposition of thermally labile, flexible segments (like polydimethylsiloxane (PDMS)) acts as a self-regulating porogen (pore-forming agent), locally generating highly dispersed micropores. Alternatively, high-performance Hybrid CMS (HCMS) membranes are created by incorporating nanoparticles or Metal-Organic Framework (MOF) fillers into the polymer precursor. As depicted in Fig. 1A, this templating approach contrasts the standard polymer route by showing how carbonizable nanoparticles guide the formation of the carbon network, providing specific and enhanced separation pathways that enable these advanced CMSMs to consistently surpass the $CO_2/CH_4$ upper bound, marking the pinnacle of rational materials design [26].

## 2.2 *Advanced Pyrolysis and Activation Techniques*

The pyrolysis (or carbonization) step, which thermally transforms the polymeric precursor into the rigid, functional carbon network, remains a critical manufacturing bottleneck for Carbon Molecular Sieve Membranes (CMSMs) [27]. Consequently, current innovations are focused intensely on making this thermal process more precise, efficient, and reproducible [28]. A major area of advancement involves finely controlling the Pyrolysis Atmosphere and Rate [29]. Moving beyond simple vacuum or inert gas heating, researchers are strategically utilizing specific environments—such as pure Argon or Nitrogen, or even introducing trace amounts of reactive gases—to precisely manipulate the formation and removal of volatile byproducts [30]. This careful control directly governs the final pore size distribution and the density of the rigid carbon structure, a key factor in boosting the separation efficiency for challenging gas pairs like $N_2/CH_4$ [31].

Beyond controlling the carbonization environment, the use of Chemical and Physical Post-Treatment is increasingly common to "fine-tune" the pore network after the high-temperature synthesis [32]. Techniques such as mild oxidation, etching agents, or the application of ultrathin carbon layers via Chemical Vapor Deposition (CVD) are employed to achieve "pore throat constriction," effectively narrowing the active pore apertures [33]. This precision adjustment maximizes the molecular sieving effect, leading to superior separation performance and selectivity [34].

To address the commercial necessity of high gas throughput, innovations in fabricating Supported Thin Films are crucial [35]. Since industrial viability requires the selective carbon layer to be ultra-thin (~1 μm) for high permeance (flux), efforts are concentrating on developing robust, defect-free Thin Film Composite (TFC) carbon membranes [36]. This involves using techniques like controlled coating, vapor deposition, or *in situ* conversion onto highly porous, high-flux support materials (e.g., ceramics or specialized high-Tg polymers), thus ensuring mechanical stability while maintaining the high performance of the selective carbon layer. (The schematic of advanced membrane architecture and post-treatment techniques is shown in Fig. 2).

The innovation is also focused on making these processes scalable. For instance, Fig. 2D presents a block diagram detailing the pilot-scale production process of a carbon membrane (CM) derived from cellulose acetate hollow fibers (CAHF) [37]. This outlines the necessary industrial sequence: spinning, stabilization, and carbonization. Furthermore, research on TFC structures, such as the fabrication and formation mechanism of hybrid membranes using precursors like H-PAN/$MoS_2$/PEI

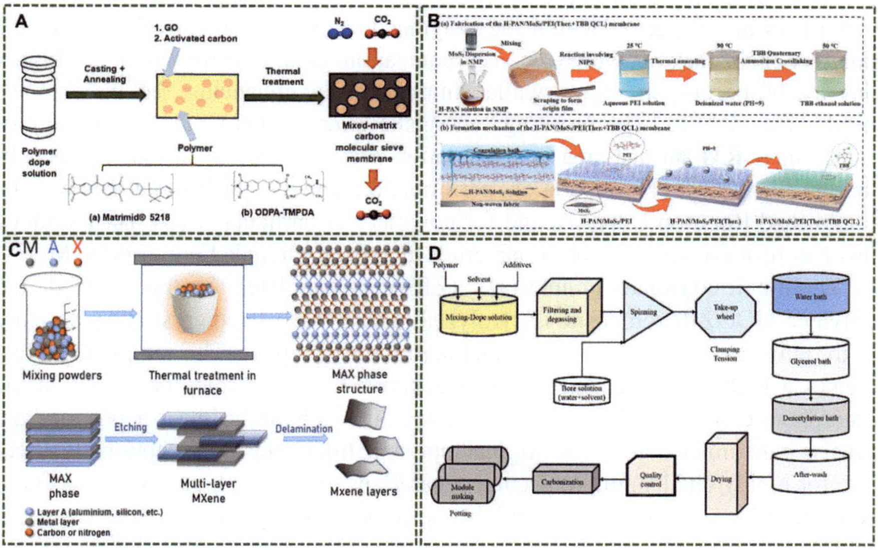


**Fig. 2** (**A**) Synthesis scheme for the development of mixed-matrix carbon molecular sieve membranes containing (1) graphene oxide (GO) and (2) activated carbon derived from: (**a**) Matrimid® 5218 and (**b**) ODPA-TMPDA polymers [39]. Reproduced from ref. [39] with permission from MDPI, copyright 2021. (**B**) Schematic illustration of (**a**) the fabrication process and (**b**) the formation mechanism of the H-PAN/$MoS_2$/PEI (Ther. + TBB QCL) membrane [38]. Reproduced from ref. [38] with permission from MDPI, copyright 2025. (**C**) Schematic diagram of MXene synthesis [40]. Reproduced from ref. [40] with permission from MDPI, copyright 2025. (**D**) Block diagram for the pilot-scale production process of carbon membrane (CM) from cellulose acetate hollow fibers (CAHF) [37]. Reproduced from ref. [37] with permission from MDPI, copyright 2018

shown in Fig. 2B, emphasizes the importance of chemical manipulation (like QCL) alongside thermal treatment to achieve desired structure and function [38]. Other advanced strategies involve integrating diverse carbon and metal materials; for example, the synthesis scheme for developing mixed-matrix CMSMs containing graphene oxide (GO) and activated carbon derived from different polymer precursors is outlined in Fig. 2A. These diverse methods highlight the convergence of material chemistry and engineering precision required to optimize the final membrane product [39].

### 2.3 *Strategic Use of Carbon Nanofillers*

A major technological leap in carbon membrane development is the creation of Mixed-Matrix Carbon Molecular Sieve Membranes (MM-CMSMs), where high-performance, carbon-based nanofillers are strategically integrated into the polymer precursor prior to pyrolysis. These nanofillers serve a dual purpose: they act both as pore-forming agents and as structural guides that modify the polymer chain packing. The incorporation of two-dimensional materials like Graphene Oxide (GO) sheets or ultra-small Carbon Dots (CDs) exemplifies this approach. When these nanofillers are dispersed within the polymer matrix, they effectively disrupt the polymer chain packing, which, upon carbonization, guides the formation of a higher density of porous structures while simultaneously enhancing the membrane's mechanical stability. The resulting synergistic effect yields CMSMs with significantly increased surface area and separation performance—such that key benchmarks, like the $H_2/N_2$ separation performance, can be pushed far beyond the established Robeson Upper Bound for polymers. A particularly advanced strategy involves using structured, porous materials like Metal-Organic Frameworks (MOFs) or Zeolitic Imidazolate Frameworks (ZIFs) as templates. In this method, the polymer-MOF/ZIF composite is pyrolyzed, where the ordered crystalline structure of the embedded framework acts as an internal hard template. This templating effect creates a highly uniform, interconnected network of ultra-micropores within the amorphous carbon matrix. Because these structures inherit the intrinsic porosity and precise pore dimensions of the original inorganic filler, they are exceptionally effective for highly challenging molecular separations, including high-efficiency $O_2/N_2$ separation for air enrichment and highly selective $CO_2$ capture.

## 3 Hybrid and Composite Material Integration for Enhanced Performance

Pure carbon membranes, while possessing superior intrinsic separation ability and chemical stability, frequently encounter practical limitations concerning processability, intrinsic mechanical brittleness, and the significantly high energy costs associated with the precursor pyrolysis. Consequently, a pivotal innovation driving the field forward is the strategic creation of hybrid and composite membrane architectures designed to exploit synergistic properties by marrying the performance of carbon materials with the flexibility and scalability of polymers.

## *3.1 The Mixed-Matrix Membrane (MMM) Strategy*

The primary manifestation of this approach is the Mixed-Matrix Membrane (MMM), where a dispersed rigid filler—either inorganic (like MOF) or carbon-based—is meticulously embedded within a continuous polymer matrix [41]. (See Fig. 3 for a general illustration of the MMM concept). This composite design cleverly aims to achieve the high selectivity characteristic of the rigid filler (often acting as molecular sieves) while retaining the flexibility, cost-effectiveness, and ease of scalable fabrication inherent to the polymer [42]. Figure 3A reinforces this point by schematically illustrating the extensive applications of both pure MOF membranes and MMMs across gas and liquid separation fields [43].

Key integrations involve utilizing highly porous crystalline materials such as Metal-Organic Frameworks (MOFs) [44] and Covalent Organic Frameworks (COFs) [45, 46]. These structures act as uniform, tunable molecular gates, drastically enhancing separation selectivity, particularly for critical gas pairs like $CO_2$/$CH_4$ and $O_2/N_2$ at ambient temperatures [47]. This enhancement often pushes the overall performance well past the theoretical polymeric Robeson upper bound. Figure 3B provides a detailed case study of this strategy in gas separation, showing the structural design and performance results of an MMM utilizing a PIM − 1 matrix. The accompanying graphs demonstrate not only the successful structural integration (panels e-h) but also the significant improvement in $CO_2/N_2$ separation factor and $CO_2$ permeability (panels i-l), culminating in performance that clearly surpasses the Robeson upper bound (panel m) [48].

## *3.2 Multifunctional Nanofillers and Operational Stability*

Furthermore, the integration of diverse nanofillers—such as Graphene Oxide (GO), carbon nanotubes (CNTs), and specialized functionalized nanoparticles—is crucial for developing multi-functional membranes [49]. For example, incorporating magnetic $Fe_3O_4$ nanoparticles into a carbon-based matrix can create membranes capable of both separation and catalytic functionality, or simply provide rapid pollutant remediation via external magnetic fields [50]. This hybrid approach is paramount in addressing real-world operational challenges [51]. MMMs are critical for improving membrane resistance to plasticization (a common failure mode in high-pressure gas separation caused by the swelling of the polymer matrix) and significantly mitigating membrane fouling in liquid separations by creating smoother, more

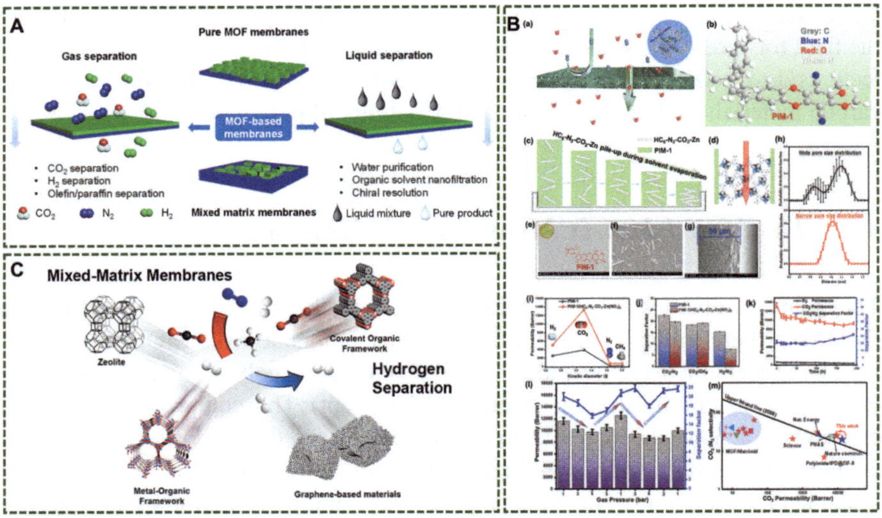


**Fig. 3** (**A**) Schematic illustration of extensive applications of pure MOF membranes and MMMs for gas and liquid separation [43]. Reproduced from ref. [43] with permission from MDPI, copyright 2023. (**B**) Mixed matrix membranes for gas separation (**a**) diagram of gas separation through a membrane. (**b**) Ladder-like molecular structure of PIM-1 (grey: C, blue: N, red: O, white: H). (**c**) Illustration of the solidification process of the stick-shaped PIM-1/$HC_6$–$N_3$–$CO_2$–$Zn(NO_3)_2$ (1%) membrane. (**d**) Schematic illustration of gas transport channels through the PIM-1/$HC_6$–$N_3$–$CO_2$–$Zn(NO_3)_2$ membrane. (**e–g**) Top view, bottom view and cross-section of the 1 wt% PIM-1/$HC_6$–$N_3$–$CO_2$–$Zn(NO_3)_2$ (1%) membrane. (**h**) Free volume hole diameters of pure PIM-1 membrane (top) and PIM-1/$HC_6$–$N_3$–$CO_2$–$Zn(NO_3)_2$ (1%) MMMs (bottom) determined by positron annihilation lifetime distributions (PALS) under vacuum at 298 K. (**i**) Single-gas separation permeability of PIM-1, PIM-1/$HC_6$–$N_3$–$CO_2$–$Zn(NO_3)_2$ (1%) membranes as a function of gas kinetic diameter at 298 K and 2 bar. (**j**) Separation factors of PIM-1 and 1 wt% PIM-1/$HC_6$–$N_3$–$CO_2$–$Zn(NO_3)_2$ (1%) membranes for $CO_2/N_2$, $CO_2/CH_4$ and $H_2/N_2$ separations. (**k**) Long-term tests of 1 wt% PIM-1/$HC_6$–$N_3$–$CO_2$–$Zn(NO_3)_2$ (1%) MMMs for $CO_2$ and $N_2$ at 298 K and 2 bar. (**l**) $CO_2$ permeability and $CO_2/N_2$separation factor of 1 wt% PIM-1/$HC_6$–$N_3$–$CO_2$–$Zn(NO_3)_2$ (1%) MMMs as a function of pressure. (**m**) A comparison of the Robeson upper bound for $CO_2/N_2$ separation between 1 wt% PIM-1/$HC_6$–$N_3$–$CO_2$–$Zn(NO_3)_2$ (1%) MMMs and literature data [48]. Reproduced from ref. [48] with permission from Royal Society of Chemistry, copyright 2024. (**C**) Recent Progress in Mixed-Matrix Membranes for Hydrogen Separation [53]. Reproduced from ref. [53] with permission from MDPI, copyright 2021

hydrophilic surfaces [52]. The continued focus on MMMs for high-demand applications, such as hydrogen separation, remains a major research thrust (Fig. 3C). Ultimately, the future of high-performance separation lies in designing robust, multi-layered architectures that intelligently utilize these hybrid materials, ensuring the high flux and long-term operational stability demanded by demanding industrial environments [53].

## 4 Artificial Intelligence and Computational Modeling in Membrane Design

The process of discovering and optimizing new high-performance carbon-based membrane materials has traditionally been a time-consuming, trial-and-error endeavor that relies heavily on extensive experimental synthesis and testing [54]. The field is now experiencing a radical transformation through the integration of Artificial Intelligence (AI) and Computational Modeling, which act as predictive tools to dramatically accelerate materials discovery and optimization [55]. On the fundamental level, Computational Modeling, including Molecular Dynamics (MD) simulations [56] and Density Functional Theory (DFT), plays a crucial role [57]. These methods allow researchers to accurately model gas or liquid transport mechanisms at the atomic scale, predicting critical properties like pore size distribution, diffusion coefficients, and solubility before any physical synthesis is performed [58]. For example, MD simulations can map how $CO_2$ or $H_2$ molecules interact with and pass through the carbon pore network, revealing the exact parameters that lead to high selectivity [59].

This highly detailed, atomic-scale data feeds into Machine Learning (ML) and AI algorithms, which constitute the next layer of innovation. ML models are trained on vast datasets of existing membrane materials, processing structure-property correlations (e.g., relating polymer precursor structure and pyrolysis conditions to final separation performance) [60, 61]. Once trained, these models can rapidly perform High-Throughput Virtual Screening, evaluating thousands of novel material candidates (such as potential new polymer precursors or MOF fillers) in a fraction of the time required by experimental methods [62]. The general strategy involves using data to train a model that predicts performance, which then informs the search for new materials. Figure 4B presents a schematic of an ML workflow specifically within the data-driven innovative material discovery process. Furthermore, the complexity of these workflows can be tailored to specific material classes; for instance, Fig. 4A outlines dedicated ML workflows designed for the efficient optimization and screening of Organic Framework Membranes (OFMs).

This approach allows researchers to efficiently target only the most promising materials that are predicted to surpass the current performance upper bound, thereby minimizing synthesis costs and research cycles. Ultimately, the synergy between precise atomic-scale modeling (DFT/MD) and data-driven prediction (AI/ML) is establishing a new paradigm of "design-before-synthesis," which is essential for developing the next generation of highly optimized, purpose-built carbon membranes.

## 5 Prospects and Challenges in Commercialization

The scientific superiority of carbon-based membranes, demonstrated by their ability to achieve high selectivity and flux simultaneously, fuels significant optimism regarding their industrial future [69]. However, the path from laboratory

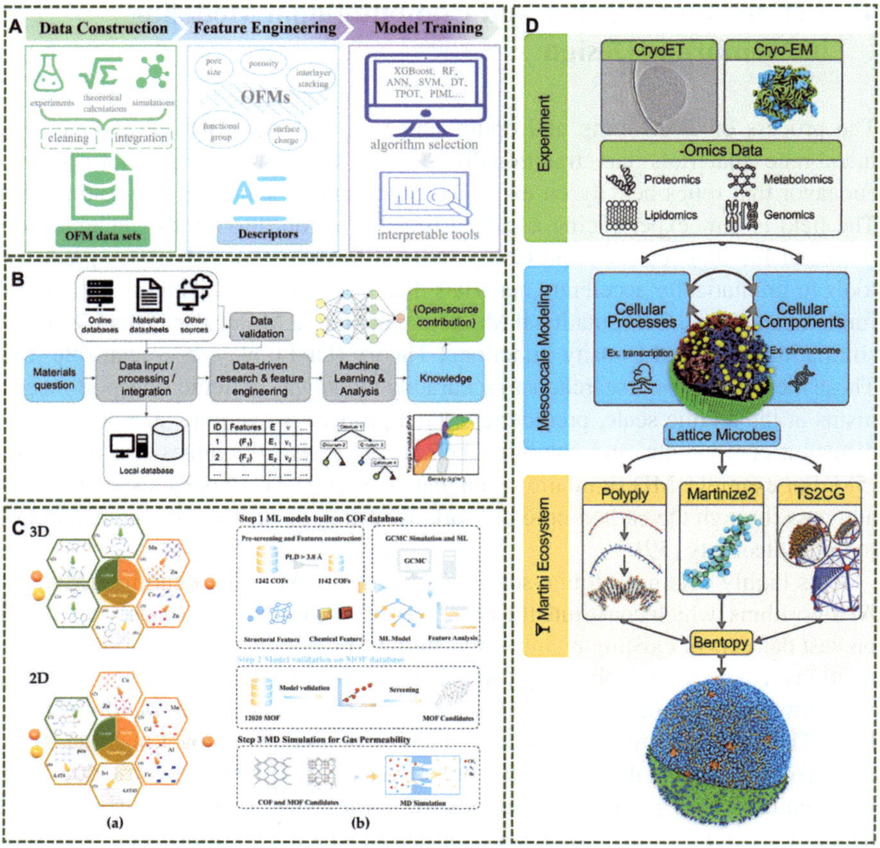


**Fig. 4** (**A**) The workflows for machine learning (ML) in organic framework membrane (OFM) design [63]. Reproduced with permission [63]. Copyright 2025, MDPI. (**B**) Schematic of an ML workflow in the data-driven innovative material discovery process. Reproduced with permission [64]. Copyright 2020, American Chemical Society. (**C**) (**a**) Design strategies of 3D and 2D MOF membranes for boosting Kr/Xe separation. Reprinted with copyright permission from Ref. [54] Copyright 2019 Elsevier Ltd.; (**b**) a workflow for developing universal machine learning models to screen high-performance COFs/MOFs membrane materials for helium separation. Reprinted with copyright permission from Ref. [65] Copyright 2024 Elsevier Ltd. (**D**) Integrative modeling workflow for building *in silico* whole-cell models. The initial step consists of collecting experimental data to inform the *in silico* modeling. Data from CryoET images (Image from EMD-23661 by Lam and Villa ref. [66]), Cryo-EM protein structures and -Omics experiments are incorporated into our cell models. The second stage in the workflow concerns mesoscale modeling. Here a kinetic model of the whole JCVI-syn3A (Image ref. [67]) is used to gain quantitative insights into cellular processes and composition. As part of the final step, Martini models of the cellular components are generated using tools in the Martini ecosystem: Polyply, Martinize2, and TS2CG (Image ref. [68]). Lastly, Bentopy facilitates the assembly of the individual molecular components in their appropriate abundances into the final molecular-resolution whole-cell model

breakthrough to full commercialization is fraught with technological and economic challenges that must be systematically addressed [70]. The single greatest hurdle is Scalability and Cost. Manufacturing carbon membranes typically involves high-temperature pyrolysis and the use of specialized, often expensive, polymer precursors and inert gases [71]. To compete with established, low-cost polymeric membranes (like cellulose acetate), researchers must pioneer low-cost, high-throughput fabrication methods, such as continuous roll-to-roll processing or advanced hollow fiber spinning and carbonization techniques, to drastically reduce the capital expenditure (CAPEX) associated with membrane modules [6]. Equally critical is ensuring Long-Term Operational Stability. While carbon membranes are chemically and thermally resistant, they can suffer from mechanical brittleness and performance degradation due to long-term exposure to contaminants [72]. For high-pressure gas separation, preventing membrane failure due to plasticization (the swelling and pore widening caused by high partial pressures of sorbing gases like $CO_2$) [73] and mitigating fouling (the deposition of particulates or organics in liquid separation) remain intensive areas of focus, requiring the development of robust, supported thin-film composite architectures [74]. Finally, Process Reproducibility and Quality Control are non-negotiable for industrial adoption. Current CMSM synthesis methods often yield variable results, making quality control a major headache [75]. Future efforts must focus on developing standardized protocols and *in situ* monitoring systems—potentially leveraging the AI models discussed in the previous section—to guarantee that every manufactured membrane module meets stringent and consistent performance specifications, thereby overcoming investor and industry reluctance towards integrating a complex, nascent technology.

## 6 Conclusion and Future Research Directions

This chapter has established that carbon-based membrane technologies (CMTs) represent a transformational departure from conventional separation methods, offering a pathway toward energy-efficient and highly selective separationacross environmental and industrial applications. Breakthroughs in rational precursor design (Sect. 2), the synergistic integration of hybrid materials like MOFs and nanofillers (Sect. 3), and the disruptive power of AI and computational modeling (Sect. 4) are collectively propelling CMSMs past the performance limitations of their polymeric counterparts. These innovations promise to revolutionize gas purification, $CO_2$ capture, and high-temperature industrial separations. However, realizing the immense commercial prospects (Sect. 5) requires a deliberate shift in research focus from laboratory-scale performance enhancement to overcoming core engineering challenges. Moving forward, the most critical future research directions must concentrate on: (1) Scale-up and Manufacturing:Developing sustainable, continuous, and low-cost processes (such as specialized hollow fiber pyrolysis) to enable mass production of defect-free, ultra-thin selective layers. (2) Long-Term Operational Integrity: Enhancing mechanical robustness, specifically resistance to plasticization in high-pressure environments, and developing anti-fouling strategies that ensure stable performance over thousands of operational hours. (3) Predictive Digitalization: Fully integrating advanced machine learning models with experimental robotic platforms to create closed-loop systems that can automatically refine synthesis parameters and rapidly discover new optimal precursors and filler combinations. By tackling these challenges through a concerted effort spanning materials science, chemical engineering, and data science, CMTs are poised to transition from a leading-edge research topic to a cornerstone technology for a sustainable industrial future.

## References


1. Noble, R.D., Agrawal, R.: Separations research needs for the 21st century. Ind. Eng. Chem. Res. **44**(9), 2887–2892 (2005)
2. Rethinasabapathy, M., et al.: Recent progress in functional nanomaterials towards the storage, separation, and removal of tritium. Adv. Mater. **35**(48), 2301589 (2023)
3. Regmi, C., Kshetri, Y.K., Wickramasinghe, S.R.: Carbon-based nanocomposite membranes for membrane distillation: progress, problems and future prospects. Membranes. **14**(7), 160 (2024)
4. Llosa Tanco, M.A., Pacheco Tanaka, D.A.: Recent advances on carbon molecular sieve membranes (CMSMs) and reactors. Processes. **4**(3), 29 (2016)
5. Alkandari, S.H.: Advanced and sustainable manufacturing methods of polymeric based membranes for gas separation (2024)
6. Mollahosseini, A., et al.: Polymeric membranes in carbon capture, utilization, and storage: current trends and future directions in decarbonization of industrial flue gas and climate change mitigation. Energy Environ. Sci. **18**(11), 5025–5092 (2025)
7. Li, J.-R., Kuppler, R.J., Zhou, H.-C.: Selective gas adsorption and separation in metal–organic frameworks. Chem. Soc. Rev. **38**(5), 1477–1504 (2009)
8. Widakdo, J., et al.: Graphene oxide-melamine nanofilm composite membrane for efficient $CO_2$ gas separation. Sep. Purif. Technol. **323**, 124521 (2023)
9. Widakdo, J., et al.: Bioinspired ionic liquid-graphene based smart membranes with electrical tunable channels for gas separation. Appl. Mater. Today. **27**, 101441 (2022)
10. Widakdo, J., et al.: Mechanism of a self-assembling smart and electrically responsive PVDF–graphene membrane for controlled gas separation. ACS Appl. Mater. Interfaces. **12**(27), 30915–30924 (2020)
11. Widakdo, J., et al.: Switching gas permeation through smart membranes by external stimuli: a review. J. Mater. Chem. A. **10**(32), 16743–16760 (2022)
12. Widakdo, J., et al.: Tailoring of graphene–organic frameworks membrane to enable reversed electrical-switchable permselectivity in $CO_2$ separation. Carbon. **182**, 545–558 (2021)
13. Marchetti, P., Peeva, L., Livingston, A.: The selectivity challenge in organic solvent nanofiltration: membrane and process solutions. Annu. Rev. Chem. Biomol. Eng. **8**(1), 473–497 (2017)
14. Subrahmanya, T., et al.: Effect of functionalized nanodiamonds and surfactants mediation on the nanofiltration performance of polyamide thin-film nanocomposite membranes. Desalination. **555**, 116540 (2023)
15. Widakdo, J., et al.: Positron annihilation spectroscopy for the free volume depth profile analysis of multilayer and 2D materials composite membranes: a review. Sep. Purif. Technol. **322**, 124366 (2023)
16. Abebe, S.H., et al.: Zwitterionic M-PhA modified GO nanocomposite membranes for enhanced organic solvent nanofiltration and wastewater treatment with antifouling performance. J. Membr. Sci. **736**, 124596 (2025)

17. Seftor, E., et al.: Application of chemically modified tetracyclines (CMTs) in experimental models of cancer and arthritis. Adv. Dent. Res. **12**(1), 103–110 (1998)
18. Comesaña-Gándara, B., et al.: Gas separation membranes made through thermal rearragement of ortho-methoxypolyimides. In: Influence of the Ortho Substitution on the Final Properties
19. Farid, M.U., et al.: Advancements in nanoenabled membrane distillation for a sustainable water-energy-environment nexus. Adv. Mater. **36**(17), 2307950 (2024)
20. Lucas, R., et al.: DOE Advanced Scientific Computing Advisory Subcommittee (ASCAC) Report: Top Ten Exascale Research Challenges. USDOE Office of Science (SC)(United States) (2014)
21. Haider, S.: A semi-industrial scale process to produce carbon membranes for gas separation (2019)
22. Rungta, M., et al.: Carbon molecular sieve structure development and membrane performance relationships. Carbon. **115**, 237–248 (2017)
23. Liang, J., et al.: Effects on carbon molecular sieve membrane properties for a precursor polyimide with simultaneous flatness and contortion in the repeat unit. ChemSusChem. **13**(20), 5531–5538 (2020)
24. Qiu, W., et al.: Key features of polyimide-derived carbon molecular sieves. Angew. Chem. **133**(41), 22496–22505 (2021)
25. Hazazi, K., et al.: Precise molecular sieving of ethylene from ethane using triptycene-derived submicroporous carbon membranes. Nat. Mater. **22**(10), 1218–1226 (2023)
26. Lee, T.H., et al.: In situ derived hybrid carbon molecular sieve membranes with tailored ultramicroporosity for efficient gas separation. Small. **17**(47), 2104698 (2021)
27. Lin, Y.-T., et al.: Orientation of the microstructure of free-standing carbon hollow fiber membranes modulated by carbonization condition triggers aging-resistant properties. J. Environ. Chem. Eng. **10**(6), 108941 (2022)
28. Tofail, S.A., et al.: Additive manufacturing: scientific and technological challenges, market uptake and opportunities. Mater. Today. **21**(1), 22–37 (2018)
29. Ibrahim, H.A.-H.: Introductory chapter: pyrolysis. Recent Adv. Pyrolysis, 1 (2020)
30. Rahimpour, M.R., Makarem, M.A., Meshksar, M.: Advances in Natural Gas: Formation, Processing, and Applications. Vol. 5: Natural Gas Impurities and Condensate Removal. Elsevier (2024)
31. Hosseini, S.S., Chung, T.S.: Carbon membranes from blends of PBI and polyimides for $N_2/CH_4$ and $CO_2/CH_4$ separation and hydrogen purification. J. Membr. Sci. **328**(1–2), 174–185 (2009)
32. De, S., et al.: Biomass-derived porous carbon materials: synthesis and catalytic applications. ChemCatChem. **7**(11), 1608–1629 (2015)
33. Creighton, J., Ho, P.: Introduction to chemical vapor deposition (CVD). ASM Int., 407 (2001)
34. Wang, H., et al.: Organic molecular sieve membranes for chemical separations. Chem. Soc. Rev. **50**(9), 5468–5516 (2021)
35. Liu, J., Wöll, C.: Surface-supported metal–organic framework thin films: fabrication methods, applications, and challenges. Chem. Soc. Rev. **46**(19), 5730–5770 (2017)
36. Saleh, E.A.M., et al.: Recent advances in thin film composite (TFC) membranes development: materials and modification methods. Environ. Sci.: Water Res. Technol. **11**, 1059–1085 (2025)
37. Haider, S., et al.: Pilot–scale production of carbon hollow fiber membranes from regenerated cellulose precursor-Part II: carbonization procedure. Membranes. **8**(4), 97 (2018)
38. Zhang, Y., et al.: Double-crosslinked H-PAN/$MoS_2$/PEI composite nanofiltration membrane for ethanol systems: fabrication and dye separation performance. Membranes. **15**(10), 286 (2025)
39. Chuah, C.Y., et al.: Carbon molecular sieve membranes comprising graphene oxides and porous carbon for $CO_2/N_2$ separation. Membranes. **11**(4), 284 (2021)
40. Suleimenova, M., et al.: Composite membranes based on mxene and nanocellulose for water purification: structure, efficiency, and future prospects. Membranes. **15**(10), 293 (2025)
41. Arundhathi, B., et al.: Advancements in mixed-matrix membranes for various separation applications: state of the art and future prospects. Membranes. **14**(11), 224 (2024)

42. Zhang, W., et al.: Solution processable metal–organic frameworks: synthesis strategy and applications. Adv. Mater. **37**(1), 2412708 (2025)
43. Duan, Y., et al.: Engineering metal–organic–framework (MOF)-based membranes for gas and liquid separation. Membranes. **13**(5), 480 (2023)
44. Cai, G., et al.: Metal–organic framework-based hierarchically porous materials: synthesis and applications. Chem. Rev. **121**(20), 12278–12326 (2021)
45. Chen, Z., et al.: Hybrid porous crystalline materials from metal organic frameworks and covalent organic frameworks. Adv. Sci. **8**(20), 2101883 (2021)
46. Huang, N., Wang, P., Jiang, D.: Covalent organic frameworks: a materials platform for structural and functional designs. Nat. Rev. Mater. **1**(10), 1–19 (2016)
47. Chuah, C.Y.: Microporous Materials with Tailored Structural Properties for Enhanced Gas Separation. Nanyang Technological University (2019)
48. Du, J., et al.: Heterocycles for direct air capture and MOFs prepared from $CO_2$ utilization. J. Mater. Chem. A. **12**(13), 7711–7723 (2024)
49. Anani, J., Shokry, H., Elkady, M.: Nanofillers: design, performance, and prospects. In: Handbook of Nanofillers, pp. 313–347. Springer (2025)
50. Anjaneyulu, B., et al.: The use of magnetic porous carbon nanocomposites for the elimination of organic pollutants from wastewater. Surfaces. **7**(1), 120–142 (2024)
51. Khan, A., et al.: Comprehensive review of hybrid energy systems: challenges, applications, and optimization strategies. Energies. **18**(10), 2612 (2025)
52. Ali, S.A., et al.: Review on synthesis and characterization of advanced nanomaterials-based mixed matrix membranes (MMMs) for $CO_2$ capture: progress, challenges, and prospects. Energy Fuel. **38**(19), 18330–18366 (2024)
53. Chuah, C.Y., et al.: Recent progress in mixed-matrix membranes for hydrogen separation. Membranes. **11**(9), 666 (2021)
54. Osman, A.I., et al.: Machine learning for membrane design in energy production, gas separation, and water treatment: a review. Environ. Chem. Lett. **22**(2), 505–560 (2024)
55. Shahzad, K., Mardare, A.I., Hassel, A.W.: Accelerating materials discovery: combinatorial synthesis, high-throughput characterization, and computational advances. Sci. Technol. Adv. Mater.: Meth. **4**(1), 2292486 (2024)
56. Hollingsworth, S.A., Dror, R.O.: Molecular dynamics simulation for all. Neuron. **99**(6), 1129–1143 (2018)
57. Dawson, W., et al.: Density functional theory calculations of large systems: interplay between fragments, observables, and computational complexity. Wiley Interdiscip. Rev.:Comput. Mol. Sci. **12**(3), e1574 (2022)
58. Ricci, E., Minelli, M., De Angelis, M.G.: Modelling sorption and transport of gases in polymeric membranes across different scales: a review. Membranes. **12**(9), 857 (2022)
59. Anderson, R., et al.: Role of pore chemistry and topology in the $CO_2$ capture capabilities of MOFs: from molecular simulation to machine learning. Chem. Mater. **30**(18), 6325–6337 (2018)
60. Konstantopoulos, G., Koumoulos, E.P., Charitidis, C.A.: Digital innovation enabled nanomaterial manufacturing; machine learning strategies and green perspectives. Nano. **12**(15), 2646 (2022)
61. Zhao, Y., Chen, G., Liu, J.: Polymer data challenges in the AI era: bridging gaps for next-generation energy materials. arXiv preprint arXiv:2505.13494. (2025)
62. Xie, C., et al.: Machine learning approaches in polymer science: progress and fundamental for a new paradigm. SmartMat. **6**(1), e1320 (2025)
63. Wu, T., et al.: Machine learning in the design and performance prediction of organic framework membranes: methodologies, applications, and industrial prospects. Membranes. **15**(6), 178 (2025)
64. Wang, A.Y.-T., et al.: Machine learning for materials scientists: an introductory guide toward best practices. Chem. Mater. **32**(12), 4954–4965 (2020)
65. Lang, Y., et al.: A universal model for predicting helium separation in covalent organic frameworks and Metal–organic frameworks. Chem. Eng. Sci. **305**, 121175 (2025)

66. Gilbert, B.R., et al.: Generating chromosome geometries in a minimal cell from cryo-electron tomograms and chromosome conformation capture maps. Front. Mol. Biosci. **8**, 644133 (2021)
67. Thornburg, Z.R., et al.: Fundamental behaviors emerge from simulations of a living minimal cell. Cell. **185**(2), 345–360 (2022) e28
68. Pezeshkian, W., et al.: Backmapping triangulated surfaces to coarse-grained membrane models. Nat. Commun. **11**(1), 2296 (2020)
69. Richard, S.P.: Catalytic Membrane Reactors for e-chemicals production & processing, Ph.D. dissertation, Dept. Chem. Eng. and Chem., Eindhoven Univ. of Tech., Eindhoven, Netherlands, (2025)
70. Kadtke, J., Wells, I.: Policy challenges of accelerating technological change: security policy and strategy implications of parallel scientific revolutions (2014)
71. Kudapa, V.K., Mondal, S.: Hydrogen Membranes: Production, Fabrication, and Applications. CRC Press (2025)
72. Gittleman, C.S., Coms, F.D., Lai, Y.-H.: Membrane durability: physical and chemical degradation. In: Polymer Electrolyte Fuel Cell Degradation, p. 15 (2011)
73. Kadirkhan, F., et al.: Recent advances of polymeric membranes in tackling plasticization and aging for practical industrial $CO_2/CH_4$ applications—a review. Membranes. **12**(1), 71 (2022)
74. Behroozi, A.H., et al.: Membrane fabrication and modification by atomic layer deposition: processes and applications in water treatment and gas separation. ACS Appl. Mater. Interfaces. **15**(11), 13825–13843 (2023)
75. Haider, R., et al.: Engineering hyper-personalization: software challenges and brand performance in AI-driven digital marketing management: an empirical study. Int. J. Sci. Res. Arch. **15**(2), 1122–1141 (2025)



MIX
Papier aus verantwortungsvollen Quellen
Paper from responsible sources
FSC® C105338

If you have any concerns about our products,
you can contact us on
**ProductSafety@springernature.com**

In case Publisher is established outside the EU,
the EU authorized representative is:
**Springer Nature Customer Service Center GmbH**
**Europaplatz 3, 69115 Heidelberg, Germany**

Printed by Libri Plureos GmbH
in Hamburg, Germany